I0108576

A Scientist's Approach to Honest Parenting

By Cedric O. Buckley, Ph.D.
Micah Murdock, Contributor

Real Issues Publishing
Dallas, Texas

Cover Design: Image attrition to Canva®

A Scientist's Approach to Honest Parenting

Copyright ©

ISBN: 979-8-218-10398-9

All rights reserved. No part of this book may be reproduced, stored in retrieval system, or transmitted by any means – electronic, mechanical, photographic (photocopying), recording, or otherwise – without prior permission in writing from the author – unless the author's name, blurb, sole credit and/or contact information is attached. For more information, contact realissuespublishing@gmail.com.

Printed in USA
Real Issues Publishing
Dallas, TX 75228

"Clearly, God is trying to lead us towards perfection as individuals, but God is also perfecting us in our roles with one another. We often put unrealistic expectations on ourselves as parents by saying, "I must be the perfect parent now," when the truth is that there is no such thing as a perfect parent. However, there is a parent being perfected by God and striving to follow the example of God as parent. I think this is a very good theological point concerning God as a parent and very encouraging to the new mother and the new father. Also, I think it speaks to the individual who is not married, yet and not a parent, yet. In my opinion, it is a point that is intergenerational and intersectional for every listener."

- Rocky Shack, Husband, Father,
Missions & Community Outreach Minister

"[The] behavior modeling - great scenarios and practical usage!"

"This is a great effort to assist new parents, and especially many millennials, to navigate the numerous landmines of parenting. Pat yourself on the back for taking this on. Parenting is noble work, and trying to provide guidance for this most important of lifetime duties is no small undertaking. Congratulations!"

- Leo Pinson, Solutions Consultant

"It's almost as if you read my journal. I also thought that the follow-up questions [are very] thought-provoking [as they] caused me to look at myself."

– Kelcie Shack, Wife & Mom

TABLE OF CONTENTS

Acknowledgements 9

Introduction.. .11

Chapter 1: Identify Yourself With the "Ultimate Parent".....**21**

Chapter 2: Observe Who They Are Becoming.......................**31**

Chapter 3: Bring Consistency into Conflicts...........................**51**

Chapter 4: An Honest Approach to Learning.................................**67**

Chapter 5: Check Your Tech and Other Parental Safeguards........**77**

Chapter 6: Parenting Through COVID-19................................**91**

Epilogue: Keep Connected With Honest Parents.................**97**

For my sons, Corland and Nicholas. -Love, Dad.

.

Acknowledgements

This book likely would not have happened but for a nasty virus called SARS-CoV-2. Untold global suffering and loss along with unanswered questions have been left in its wake. It seems fitting that I acknowledge this global pandemic, but not for the horrible devastation it wrought. I want to acknowledge all of us who made it through! Though losses are countless and lives are changed forever, out of grief and suffering we were able to find new beginnings. Collectively, we found the life force within and harnessed its transformative power. Some of us found a new place to call home. Some of us found a new career. Some of us found a new love. So here's to each of you reading this…you made some great lemonade out of the lemons you were given!! Never forget how grateful we all are to each other for doing the hard work of picking up the pieces while advancing our lives and society.

Thank you to several early reviewers of various drafts of this book, including Pastor Rocky Shack and Mrs. Kelcie Shack. Thanks also to my dear friends Leo and Elaine Pinson. To all who will read *Honest Parenting* and upcoming installments from the "A Scientist's Approach" series, I sincerely thank you for your support. May your curiosity continue to provide food for your soul. Know that for me, science is not only about being willing to question, but also being willing to listen for the answers and respond in love and gratitude. Life becomes much richer for each of us as we spend more and more of our energies focused not on the destination, but on the journey. To my Duke Ellington School of the Arts family. I am grateful for the opportunity to help guide and nurture you, the next generation of artist/scholars. Don't be deterred by life's twists and turns. Don't give in. Never give up. We are counting on you. Make us proud!

Special thanks to my editor, Staci Sweet, Staci Sweet & Company, and Real Issues Publishing. In a world that can seem full of disappointments, it is so rewarding to work with such a professional, dedicated group. I also thank my family, especially my mother (Mrs. Myra B. Buckley) and my father (Rev. Horace L. Buckley). Your love, guidance and nurturing throughout my life has produced a son who truly honors and respects you both as parents. You gifted the world four children and gave us all you had. Thank you for all you sacrificed for us. I'm still learning to be a better parent, spouse and friend through your examples.

To my wife, Tonya Holland-Buckley. What can I say? I am forever grateful you picked up my phone call! Our life together continues to grow me in ways I never imagined. Your love and support make all the difference. May our expressions of love for each other provide a continuously positive example and source of inspiration for our sons.

**Practice
(as with medicine or in any other science)
is Required!**

As parents, we recognize we have an awesome responsibility to our children. We want to get this right. Everywhere we turn, we see tragic situations and circumstances involving children that worry us. We wonder if we are doing what we need to be doing to make certain our children won't be among those we see in news reports: uninterested in reading, unwilling to follow rules of "acceptable" behavior, choosing to hide things from us, getting in trouble with law enforcement and the judicial system, becoming violent, experimenting with illicit drugs and sex, and the list goes on. As parents, we know where these types of behaviors can land our children and we desperately want answers, and to have plans in place that will guarantee our children won't become the next statistic in someone's report. We want our children to grow up safe, healthy, educated and ready to go after their dreams. Well, I am here to tell you that those wishes we have for our children are not unreasonable...and they are quite attainable.

To be quite **honest**, I wrote this book because I am that parent. I share those same dreams and fears. I have worried. I have prayed. I have cried. In the end, I have come to realize that as parents, we have a way forward. We do not have to live in the fear that we somehow won't raise our children the "right way". We can move forward using many of the methods I learned through years of education and training in the world of scientific research. An interesting thing I discovered for myself during those years of research training is this:

Just because an *action* taken appears simple, does not mean that the *responses* to that action will be simple.

In mathematics there is an algebraic property that reads: if A=B and B=C, then A=C. It turns out that this *transitive property of equality* is not always directly applicable in human systems; therefore, for human systems, I learned instead: if A=B and B=C, then A may or may not = B OR C!

In other words, many human actions, by us as parents, that result in the behaviors our children go on to learn and display are not linear. Meaning, we usually are not able to pinpoint the actions taken that directly resulted in the learned behaviors of our children. In addition, we are not able to raise our children within the controlled laboratory environments with which I am quite familiar. You see, in laboratory settings we can tightly and specifically control the environment so that in some ways, we have a measure of control over the range of possible outcomes; not so in parenting. As parents, all our efforts are competing against every other imaginable input that may find its way to our precious children. And so, we react as any parent would. We attempt to erect barriers in a futile attempt to control the environment in which we raise our children. Of course, it seems to work...but only for a while. Sure, we have an enormous measure of control in those early parenting years. But then we realize that our children may not have really been learning the lessons we had been teaching. We begin to wonder if, just maybe, they have simply been showing us what we want to see until they are able to make their own choices. It is when our children begin to realize they can make their own choices (from an innocuous decision like which flavor of yogurt is their favorite, to divulging to us that they are being bullied, or have experienced teen dating violence) AND start exercising their autonomy, we parents fear all hell will break loose in the home. So, is there any way to avoid this worrisome situation?

This book is my contribution. Consider it your personal, pre-emptive strike against the daily onslaught of external attacks on the practice of parenting. I am certain *a more scientific approach to honest parenting* will work for you too. To be quite honest, I use my training as a scientist to inform as well as constantly and consistently improve my own parenting. I now offer these same *scientific approaches* and principles to each of you. However, I realize that the mention of the word science, scientific or scientist causes many people instinctively to think: "correct answers", "difficult to understand", and/or "smart people." Contrary to this perception, at its core, the practice of science is founded on the principle of ordered processes. The creation and implementation of ordered processes is a universal phenomenon invoked since the beginning of time. The universe and everything in it, including you and me, exists and operates on this founding principle. Once you truly understand this, neither science nor *scientific approaches* will cause you apprehension; and, if you will allow yourself to accept the truth of this premise, it can help move you towards what I call "*honest parenting*." This is my purpose as you're reading these pages.

As a scientist, I chose to search for parenting answers using the same *scientific approach* I apply to very complex biological questions. We scientists call it the **reductionist method**. After recognizing the complexity of the problem, I reduce or split it up into more manageable sub-components.

Complex biological question

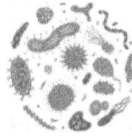

Reduce problem into manageable sub-components

I was then able to derive some key insights that I have used to guide me through raising my children. These methods and insights have a dynamic property which we refer to in science as being "**fluid**". This means as our children grow and as we as parents grow, the insights and methods will change. This is an important and often overlooked point. Because we are dealing with complex human interactions among multiple people of differing age groups, parenting methods must constantly be chosen from among a variety of methods in the parenting "toolkit". This may seem obvious on paper, but when you are in the thick of parenting, the thought of switching methods doesn't always immediately come to mind. One reason for this is that we as parents are not always tuned in to see the signs that our children now need a different *parenting approach*. Things were going along just fine, so we just kept on doing what had been working. As a scientist and a parent, I have learned that children almost always provide some clues when they need us to step outside our zones of comfort and employ a different approach.

As a responsible scientist and author, as well as a parent committed to the daily practice of these approaches, I must be upfront and present this disclaimer before we dive into the "good stuff".

Disclaimer:

My book will likely not address everything you're going to face as you move towards honest parenting.

Yes, you read that correctly. No book can do that, no matter what the authors tell you. It's simply not possible.

What this book can and will do is provide you the beginnings of the road map you will need. I say beginnings because *honest parenting is a personal, life-long journey between you, your children and God.* My role here, In this first book, is simply to jump start the journey for some, provide redirection for many (hopefully), and for a few, offer a much-needed challenge to keep moving forward toward the parenting goals you have set. Along the way, I will introduce and explain key support tools you will find useful on your unique parenting journey.

Regardless of where you may start, my hope for every parent who reads this book is for you to be successful in your parenting journey. In fact, I NEED you to be successful for my own selfish reasons. To be quite honest, I want my own children to inherit a world they can understand and successfully navigate without fear...a world that feels honest and familiar to them. So, if or when my children ever meet your children, I want there to be no confusion about the correct ways to interact with other human beings. In order for that hope to be actualized, I believe every child needs to be raised in a way that affirms the dignity and worth of themselves and every other human being; and it is my unequivocal conviction that **honest parenting** guides each of our children towards those truths about each of us as members of the human race.

As we practice **honest parenting**, we can confidently and proudly release them into this world - their world - knowing that they will prosper, they will achieve, they will overcome personal adversity, they will help others, they will support community and they will honor humanity. As **honest parents**, this is all we ever truly could want for our children, and for this world over, which God has appointed us stewards. Yet, make no mistake. If I can become an honest parent, then so can you and every other parent who will read this book. How can I make such a blanket and universally applicable claim? Two reasons. First of all, I believe each child is gifted uniquely to a set of "creators" by the Creator; secondly, either (or both) of those "creators" can accept their unique gift as confirmation of their worth, capacity, and commitment to ensure not only the survival, but the preparation of that child for human flourishing.

So, if you're looking for an excuse to forgo the customary insecurities, guilt, inadequacies, and frustrations attendant with serious parenting, surely you have purchased the right text! From the very beginning, I intend to persuade each of you that you are not a bad parent. I need to be sure I make this point clearly, because I am going to insist that you adopt this opinion and mindset right now. Too many parents worry incessantly about what they are (or are not) doing for their children. I can totally relate to those all too familiar, self-tormenting thoughts and internal strife!!!

- ✓ Yes, there are decisions to be made.
- ✓ Yes, there are repercussions for misguided thinking or misguided actions.
- ✓ Yes, the world constantly judges us relative to the choices we, as parents, make for our children...but...

As a professionally trained scientist (my Ph.D. is in Microbiology and Molecular Genetics), I have spent all my formal education, teaching and research careers studying processes. I have been blessed to convert that expertise and apply it to parenting. In so doing, I am offering a new narrative about parenting, especially for first timers and millennials. I am particularly interested in having this conversation with these two populations because I suspect that compared to any other parenting generation, first-timers and millennials spend more time doing online research, asking advice from seasoned parents, attending seminars, reading blog after blog, and yes, praying, praying...and praying. Guess what? That tells me two things: (1) you have a sincere understanding that you are a steward who has been given charge of God's most precious Creation. And, (2) you genuinely accept that you are uniquely devoted to your children and your family unit. Now while there are no perfect parents just like there are no perfect people, I do urge you to consider that the practice of parenting your children can become more perfect if you approach it as a series of ever-evolving, fluid, and progressive experiences within one ordered process.

Summarily, as you explore the pages of this book, I wholeheartedly believe that each of us can experience "*honest parenting*" because the word "parent" is both a verb AND a noun, i.e. carrying out the responsibilities of child-rearing and a person who is responsible for those collective actions. Taking into account both the noun and the verb, the "*scientific*" premise of my book is that in becoming an *honest parent* as well as in the long-term commitment to act responsibly as such, all of us will become more successful as we yield ourselves to what is an on-going, progressive and ordered process. Please keep reading, my fellow "scientists" as we enter the laboratory of life!

In the chapter it was mentioned, "Just because an action taken appears simple, does not mean that the responses to that action will be simple." In mathematics there is an algebraic property that reads:

If A=B and B=C, then A=C. It turns out that this transitive property of equality is not always directly applicable in human systems; therefore, for human systems, I learned instead: if A=B and B=C, then A may or may not = B OR C!

  a. Can you think of one or two examples in parenting your children when you witnessed "inequality," i.e. when a "simple" action on YOUR part resulted in a very complex, confusing or complicated responses from your child?

  b. How about an instance where one of your children's "simple" actions resulted in not so simple responses from you?

Recall the following and answer below.

> "However, I realize that the mention of the word science, scientific or scientist causes many people to instinctively think: "correct answers", "difficult to understand", and/or "smart people." Contrary to this perception, at its core, the practice of science is founded on the principle of ordered processes. The creation and implementation of ordered processes is a universal phenomenon invoked since the beginning of time. The universe and everything in it, including you and me, exists and operates on this founding principle. Once you truly understand this, neither science nor scientific approaches will cause you apprehension; and, if you will allow yourself to accept the truth of this premise, it can help move you towards what I call "honest parenting." This is my purpose as you're reading these pages."

a. Be honest, do you think parents who are intimidated or reluctant about scientific processes can raise children who are STEM-friendly, naturally inquisitive, and confident learners when it comes to science? Why or why not?

b. Do you think that children whose parents can afford to purchase expensive tech toys and/or send them to expensive science camps are trying to earn some sort of "right of access" relative not only to the tools for learning science, but for the purpose of steering their children into understanding science as an economic enterprise? Why or why not?

::: Chapter 1 :::
Identify Yourself With the "Ultimate Parent"

In addition to being an ordered process, **honest parenting is also a spiritual process.** Whether you have grown up within a religious framework or not, this is an inescapable truth. As a spiritual process, honest parenting is a product of our humanity, an acknowledgment that we, as humans (i.e. mere mortals) actually did not create this world into which we were born. Given this truth, we all face (whether willingly or unwillingly) the realization that there are processes in each of our lives over which we have no control. We see examples of our lack of control all around us: day turns to night and back to day again; the winds blow; the rains come; birds take flight; cows graze; fish spawn; spiders spin their intricate webs; moles burrow tunnels underneath the earth (including our well-manicured lawns) even while that once beautiful fence around the backyard rots, decays and moves into a state of disrepair; ants find their way to crumbs unintentionally left behind after a birthday party as friends and family who share in those annual celebrations live and die just like seasons change.

Approaching this inevitability of change can encourage personal enlightenment, provoke public anxiety while simultaneously ushering in a powerful and humbling spirit. These are natural responses to a process at work in each of our lives, for all of our lives. We have to learn to live in harmony with change. Every attempt to fight the process leads only to confusion and disappointment. When we fight change, we are working against God's power to heal. When we instead relinquish ourselves to change, we relinquish ourselves to God. At that point, we have harnessed the template, the "game plan", for practicing honest parenting: *a willingness to relinquish our instinctive desire to control.*

Once we agree with God that our desire to control stands in the way of the practice of **honest parenting**, we then begin to participate in the ordered, spiritual process more effectively. I know this seems counterintuitive but by letting go, we actually gain. Yes, it's true! We become open to allowing the power of The Divine to literally direct our thoughts and control our actions. We become sensitive not only to our relationship with God, but also to our relationship with our children. Now we can see them, but with God's eyes. We can hear their frustrations, but through God's ears. We can partner in their learning, growth and development, but using God's hands. We begin to parent honestly, because we have stepped out of the way and allowed God to remove our human desire for control. Now we can approach our children with the same childlike sense of wonder and amazement for life that they bring to every situation they encounter. We can be present and in the moment as they naturally reveal themselves to us day after day, hour after hour, moment after glorious moment. They trust us...because we have finally let go!

By insisting that honest parenting is a spiritual process, I do not intend to engage in an expansive discussion about the merits of religion; nor am I proposing any ability or desire to place one religion above another. Those are discussions for someone other than me, at a time other than now, in a place other than here. Whatever your religious path, I do offer you this: the reality of The Divine does not need anyone's affirmation...certainly not mine. For that reason, I can assure you that honest parenting is a spiritual process without forcing particular beliefs upon you or dictating what you ought to believe. I cannot offer you a faith formula. I can only tell you, however, that the practice of **honest parenting**, which we explore in this book, is rooted in a belief in The Divine. For me, without this rootedness, i.e., foundation, it is difficult to work expectantly towards many of the parenting goals we all seek to achieve. For me, there simply is no other option other than to approach **honest parenting** as a spiritual process. The truth is that I have tried living other ways besides or beyond a life of religious faith. None of them have worked for me to be quite honest. So, *as for me and my house, we will serve the Lord* (Joshua 24:15).

I suppose you could argue I have an advantage. I grew up in a religious household. In fact, I am the son of a Baptist minister. Until his retirement, my father pastored one of the largest, Black Baptist churches in my hometown; and, so my siblings and I grew up "in the church." Although as an adult, I explored many different religious doctrines, I have settled in my mind, body and spirit that the Christian faith is where I rest my hope. In your own spiritual practice of **honest parenting**, you too will have to make your own choice. It is the most important decision of your life here on earth. The decision you make as to whether or not you will submit to a Higher Power to govern your life will determine your ability to be effective with practicing **honest parenting**. Full stop. A personal example here may be of use.

I take my kids to the local children's museum periodically. I had never really given the outing much thought as it pertained to practicing **honest parenting**. We always had a good time, so there was no need to do anything differently, right? Well, our most recent trip started off the same as any of our previous trips.

First you lay the ground rules: "no running (they'll run anyway), watch out for other kids (kids bump into each other dad, get over it), stay where I can see you (right dad...aren't you watching us?)". Then, it's tickets, wristbands and they're off!! I make a mental note of the time and begin the countdown to wrapping up the fun and heading home. But this time was different. God had a different "game plan" for me. It was simple, yet immensely effective.

Somehow, God opened my eyes to see the wonderment in theirs as they approached and interacted with the exhibits. My job became to let go. Stop trying to control how long to stay, which exhibit to explore, how to interact with this or with that, even which children to interact with. Instead, I just observed. It was absolutely amazing to watch them! The excitement when THEY "figured it out"; the curiosity and timidity with which they approached unknown situations; the opportunities they boldly took to make friends with other children they didn't know in order to play and explore together; their periodic need to look around for me...not necessarily to ask for my help...just for reassurance that I was nearby and whatever they were currently engaged in was ok with me. This is the type of relationship with our children that promotes **honest parenting**. Because now they become open books and we as parents can then use the information they provide us to make positive, impactful decisions on how best to engage and parent them honestly. Our children begin to willingly give us the information we need in order to understand them. It is in the "understanding" of our children that effective, **honest parenting** can thrive.

When we give up our natural instinct to intervene and control, we give our children the "green light" to show us exactly who they are; who they are becoming. This is golden folks. This is the core information we need in order to do our jobs effectively. Our children have a natural instinct to give us this information. They want us to know. They long to tell us...if we would simply let go, and then listen, trusting the nature of this ordered process. I need to focus your attention, however, for just a moment on the word "trust."

From a Christian (and Jewish) religious perspective, trust is only bested by love. Trust is action oriented. Trust is where we have to put in work and effort. Trust requires both knowledge and desire on our part. Thankfully, God provides us with access to both. But we have to reach out our hand and ask for them. God does not force these gifts of love on us. God waits for us to recognize what we need and acknowledge to ourselves, and to God, that we are not able to provide it for ourselves. That's when the magic can happen!! It turns out that the trust we learn to place in God, the willingness and need to trust God becomes our template for developing the sensitivities and patience we need to practice **honest parenting**. I know of no way to begin to accomplish **honest parenting** without having made a commitment to trust in God.

Here's another reason why this step is so crucial. **Honest parenting** will require admitting you don't know something. It will require having discussions with your children without having all the answers. It will require showing your weaknesses and shortcomings. It will require believing situations will resolve themselves while not knowing how or when. You see, as a scientist, "the unknown" is a welcomed and familiar situation. We design experiments with all manner of independent and dependent variables, for the express purpose of testing hypotheses: our best guesses about the nature of some observable phenomena. We then perform these experiments to test the validity of our current hypotheses, without any forgone conclusions as to the final experimental outcomes. We willingly work within the world of the unknown without fear or favor.

As it turns out, children are prewired to function quite comfortably in these situations. In fact, they often can see beyond the artificial mental barriers we adults have erected that prevent us from trusting the process when the answers are not obvious. In those moments of wrestling with the unknown, children have a unique ability to discern truth. They seem to just know. We can put on a wonderful show and convince ourselves and other adults around us that we've got the answers, but the pure honesty of a child will always call our bluff. Our children are counting on us to parent from a place of honesty. For them, it's the only framework that makes sense. So, it's easy for them to learn and grow in an environment that celebrates and infuses honest living.

Let me end this section by asking that you meditate regularly on these three key words that should guide each of us not only in *honest parenting*, but in life: **trust**, **hope** and **love**. As you follow-through with this meditation, I invite you to consider 1 Corinthians 13:4-8 which says,

"**Love is patient, love is kind. It does not envy, it does not boast, it is not proud. It does not dishonor others, it is not self-seeking, it is not easily angered, it keeps no record of wrongs. Love does not delight in evil but rejoices with the truth. It always protects, always trusts, always hopes, always perseveres. Love never fails.**"

Let's take it a step further and make this meditation real. I want you to think of a situation where you know it has been your practice to exert control. Just to be clear here, you may need to ask your partner, a close friend or a trusted relative to tell you where you exert control...you may not be aware of it, or you may not be willing to admit it to yourself. But once you have identified the situation, I want you to say very simply this:

"God, stop me from controlling this situation.
God, remove from me the desire to control this situation.
Now God, show me what I've been missing by attempting to control this situation."

Repeat this simple prayer three times, slowly. I want you to focus on the words you are saying the first time through the prayer. Focus on each word; emphasize the word(s) that seem to speak particularly to you in this moment. As you repeat this prayer the second time, I want you to feel more and more that this is truly your desire. After which, begin to ask and answer these questions:

- Why have I felt the need to control this situation?
- What will I gain by letting go?
- What might I lose if I continue trying to control this situation?

The third and final time you repeat this prayer, I want you to visualize yourself in the situation, but having no desire to control the situation...none. See yourself moving effortlessly through the situation, allowing The Divine, who knows all and loves all, to be. See yourself telling yourself,

> **"This situation is happening just as God intends, and that's enough. I don't need to add anything to nor take anything from it."**

Now, go on with your day. When the situation you prayed and meditated on eventually presents itself, remember your agreement with yourself and with God. Sit back and watch what unfolds. You have now become a personal witness to the awesome freedom that can be enjoyed when we shift our focus away from control and towards understanding. Be amazed and thank God!

Chapter 1: Deeper Dive Discussion

Which conversation would be easier for an *honest parent* of a six or seven-year old: (1) explaining where babies come from? OR (2) explaining whether or not you believe in God?

   a.   What about for an *honest parent* of an 11-year old?
   b.   Does the sex of the child matter?
   c.   What other variable(s) would an *honest parent* consider in approaching each of those conversations?

In what ways do your responses to the first question connect to the ways you were parented and/or your own childhood experiences?

Why is it important for *honest parents* to be aware of and "own" connections to our childhoods?

::: Chapter 2 :::
Observe Who They Are Becoming (and, go with)

Much of life is about choices. The sooner we can bring our children into that mindset, the sooner they will begin the process of becoming. "Becoming what", you ask? Well, whatever and whomever they have been purposed to become. You and I actually have little control over that final outcome. It has been predestined. Our role as parents is to guide them into their path and introduce them to various tools they will need in order to continue on that path toward becoming. So, here's my question.

How in the world are they going to discover that sense of purpose and discover those tools if we are not being honest in our parenting?

Yes: **Honest Parenting**. This is crucial, so let's take this in steps...similar to the Scientific Method I used every day in my research laboratory.

First, I need to briefly define a concept called "behavior modeling." In corporate America, one of the best methods for training employees, especially staff who interact with customers, is through interactive role-play or behavior modeling. Although sometimes it is challenging to recreate scenarios that accurately depict real-world situations, still many companies have come to rely on this type of training for workforce development. In this chapter, I have connected the training method of behavior modeling to scenarios I believe will lead to behavioral modification, producing what I hope is a compelling tool which you can reference to practice honest parenting.

Now, I'll set the stage with a scenario for us to work from. I'm sure we can all relate to this. Your children have made it home from school. They have had a quick snack and tackled the remainder of the day's homework. They have decided it's time for free play and you have no objection. You have been staying on top of the types of toys they have available, so you allow them to pick and choose as they wish. You also regularly check in on what they are watching on television and YouTube® as well as the types of games they play on their tablet devices. You are feeling pretty darn good about your parenting skills and your level of organization. As you begin to relax for a bit, shouting breaks out in the house. Your children have a disagreement about what to watch on YouTube® and emotions have reached a boiling point. "I just sat down for some quiet time!", you think to yourself. Immediately you can feel your anger level rising. This situation is ripe for something explosive. So, how do we choose honest parenting in this type of situation? What can we do to not only diffuse this situation, but harness the energy of the moment and channel it towards behavioral modification? Let me provide a tentative solution in which we'll have the opportunity to enact our reductionist method, discussed earlier, in which we will reduce, break down and approach this problem as we would in the laboratory.

Here's a proposed solution that embraces **honest parenting** and uses behavior modeling: You call each child (by first name) to you and away from the location of the disagreement. This first action should immediately lower the tension among the children because you have briefly removed them from the conflict environment. Next, in a calm tone, begin the process of allowing each child to explain to you the nature of the conflict. Be sure to announce that everyone will get a turn as you do not want to allow one child to interrupt the other. Once everyone has explained the conflict as they saw it, you as a parent must distill this information and say to them what you understand the conflict to be. Now the children begin to see that you are listening to them. They'll then begin to understand that you are more concerned about the underlying reason(s) for the conflict as you help them resolve the conflict without shouting at each other. At this point, you may be tempted to provide a solution. Resist that temptation. Instead, ask each child what THEY believe a reasonable solution should look like. You may likely get some unusual responses; especially if they are not used to being asked their opinion. Still, stick with it. You may also be surprised that you'll receive some very creative solutions. Continue along until a reasonable number (your discretion here) of solutions have been proposed.

Now you all need to roll up your sleeves, evaluate each solution against the next and give the children an opportunity to express their thoughts about each proposed solution. By now, the children will likely have settled on a solution by themselves. But if not, go ahead and point out strengths and weaknesses of each proposed solution. Push them to choose a solution that everyone can agree to abide by. During this process, you may see your children moving towards a solution that you would not have chosen. This is ok. *Let them choose the solution they are comfortable with and can agree to abide by.* Once they have made this decision, they have owned the conflict and the solution. Thank them for coming to an agreement, repeat the agreement reached and let them shake on it. What you have just done is facilitated behavioral modification by using a training method called behavior modeling. In fact, the method as well as the process you used navigated your children through a real-time problem. In doing so, your children were the actors in their own conflict resolution scenario!

Importantly, as you continue to employ behavior modeling and behavioral modification as an honest parenting tool, you will notice that your children will instinctively begin to solve conflicts as they arise using the corporate training skills you've taught them!

Behavioral Model
STEP I. MODEL HONEST LIVING

Let's first agree that truth is the goal what we want to achieve and that there are positive and negative indicators we can look out for to guide us toward that goal. Truth, however, has become something of a dirty word in today's U.S. culture. We all can point to example after example of untruthful adult behavior. This is nothing new. What many parents do not realize is that the degree of technology integration among today's millennials and zoomers has absolutely produced something that is new. Our children are able to access the world within the palm of their hand! Let that sink in for just a moment. This is what we as parents are fighting against...and many of us don't have a game plan for this reality. We have to understand that our parenting efforts are being diluted in ways that previous parenting generations did not have to deal with. So, today's untruthful behaviors are being circulated and amplified within our children's social circles at exponential rates. In this context, we have to use parenting methods that not only counteract examples of dishonesty, but also amplify those truthful behavior models in a similarly exponential fashion. How can we approach this task? It must be done using a deliberate, repeatable strategy. The ideas and methods we'll now take a closer look at are foundational, so they must constantly be attended to throughout the formative parenting years (which is birth through about age 6).

Now let's be clear, I cannot teach you honesty. We each are born with that moral compass, even if we didn't receive honest parenting ourselves. We all have the Divine Spirit within that nudges us towards Truth. Without truly embracing your *honest* self, you will likely find honest parenting more challenging than it should be. So, if you are having any difficulties accessing your honest self, I need to make some suggestions here for you.

Firstly, renew your relationship with God. I cannot stress this enough. We as parents are simply stewards of these divine gifts of God called children. The life's work of being a parent really starts from within. I am going to ask some things of you throughout these chapters that will require you to make honest assessments of situations. In my opinion, neither you nor I can truly decide those honest assessments. That Eternal Truth is God-ordained and God-inspired. You may be able to get by, but at some point, you will begin to experience challenges for which you will feel completely unprepared to handle. As Thomas Paine penned,

"

These are the times that try men's [and women's] souls.

As you build your personal faith walk, you can, at the same time, incorporate parenting ideas presented in this book. Please do not make the mistake of believing you have to "get yourself right first". It's the process that matters more. The destination will be finalized by your Higher Power.

Additionally, I encourage you to find and participate regularly in an activity that promotes solitude. I can virtually guarantee you that in your personal solitude time with God, you will be practicing the surrendering of control required for Honest Parenting. Here are a few examples of ways to access personal solitude: yoga, writing in a journal, walking, jogging or cycling, fishing, gardening, drawing, painting, and/or sewing. Of course, this list is in no way exhaustive. The point is to find something to do on a regular basis (at least twice per week) that will allow you to clear your mind of the day-to-day routines of living life. It doesn't have to be long... twenty to thirty minutes is plenty. During this time, you've now created a dedicated space for what is called "self-reflection". This is the space from which God speaks to us. This is where you will find your center. This is where you will learn Truth in your life and begin to live in that Truth.

As soon as you have begun the process of becoming a more honest version of yourself, you will almost automatically begin to notice things about your children. How they respond differently depending on how you approach them, what motivates them to do what you want them to do, what chores they will do without any request from you, their likes and dislikes, as well as their secret dreams, to name a few.

These are the worlds of our children we as parents need to be able to move in and out of seamlessly. When we become honest, with ourselves, we have boldly and fearlessly moved closer to the next step: **To deliberately and repetitiously replace**

untruthfulness within our children's daily living with truthful, honest living.

You will have to model this to your children. Yes, *you* will have to lead by example. This is the essence of *honest parenting*. Once you're aware of and responsive to God's direction in your life, you can then model that honest living before your children.

STEP II. MAKE NOTES OF THE CHANGES YOU SEE

As you have more and more honest interactions with your children, you will begin to notice that they will open up, share more details, and ask more questions. They will begin to display the same honest behaviors towards you that you have been faithfully modeling towards them. Please, please, please...make some sort of record of what you are observing as you continue to engage your children. In science, many of our greatest discoveries have come about *not* because we, as scientists, were so smart. But because we kept a detailed record of what we saw during our observations. Scientists, in fact, are meticulous when it comes to data collection. Many of us say it is our "bread and butter"! Again, this is foundational and cannot be overlooked. Each child is unique and will respond in unique ways. If you are deliberate early on in recognizing and documenting these responses, it will become second nature for you to determine the approach for a particular child in a particular set of circumstances. In those instances, when you find yourself struggling with how to approach a certain situation, you will have a treasure trove of solid, reliable data on that particular child which you can reference.

As arguably the most intelligent living species on our planet, the human race has relied on visualization as one of the most important mental tools for survival. Many other species see, hear, smell, taste and even feel much more sensitively than do humans...in the physical realm. We appear to have evolved, instead, to see, hear, smell, taste and feel in the spiritual realm largely through the use of mental visualization techniques. Athletes, musicians, composers, writers, and artists of all forms spend time painting mental pictures of what success looks like to them...and then they set about manifesting those successes.

"Finally, brothers [and sisters], whatever is true, whatever is noble, whatever is right, whatever is pure, whatever is lovely, whatever is admirable--if anything is excellent or praiseworthy--think about such things."

(Philippians 4:8)

I say it this way, ***"To achieve beyond what you can imagine, you have to imagine beyond what you can see!"***

Visualization is just as important a tool for scientists. Charts, graphs, tables and diagrams help us see the data we've collected in new ways, and make connections we may otherwise have overlooked. Successfully synthesizing what may appear as chaos to many is one way we, as scientists, fine-tune our abilities to imagine answers that go beyond the obvious. We regularly use visualization tools to help us spot trends and make predictions based on underlying patterns within data. As an example, let's look at a practical exercise using data from a real-world visualization image describing trends related to a topic with which we all have become way too familiar: COVID-19.

Now, now...stay with me, because this exercise has a two-fold purpose: (1) This walkthrough is meant to give you a starting framework to use whenever you need to approach reading and interpreting charts and graphs. This method will become an especially useful skill when helping with homework and projects; and, (2) As you become more comfortable with regularly analyzing and interpreting visualization tools in the physical realm, you will become more confident in using your similar abilities (some refer to them in the collective as *intuition*) in the spiritual realm, replacing the charts and graphs with spiritual vision boards that you manifest.

At first sight, deriving any useful information from this set of data may appear daunting. However, once we finish our walkthrough, some of you will think to yourself, "Hey, I think I'm getting this for the first time!!" Those are the readers I'm most interested in engaging initially. So, if reading graphs and charts for comprehension is already in your toolbox, be patient and read on...there's nuance I'll touch on briefly that could be a gamechanger for you! Have I piqued your creative interest?? Great! Let's briefly look at some COVID-19 vaccination data (2021-2022).

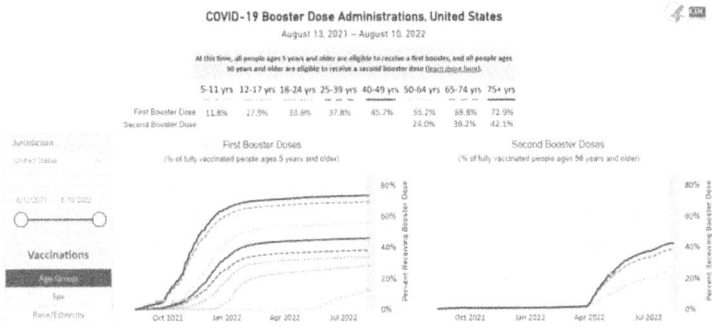

COVID-19 Booster Dose Administrations, United States
August 13, 2021 – August 10, 2022

The percentage of people who received a first booster dose includes anyone ages 5 years and older who is fully vaccinated and has received another dose of COVID-19 vaccine since August 13, 2021. This includes people who received a first booster dose and people who received an additional primary series dose as this metric does not distinguish if the recipient is immunocompromised and received an additional dose. The percentage of people who received a second booster dose includes anyone ages 50 years and older who is fully vaccinated and has received two subsequent does of COVID-19 since August 13, 2021. This includes people who received two booster doses and people who received one additional dose and one booster dose. Due to the time between vaccine administration and when records are reported to the CDC weekly, vaccinations administered during the last 6 days may not yet be reported. This reporting log is represented by the gray, shaded box.

Last Updated: August 10, 2022
Data source: VTrcks, Iis, Federal Pharmacy Program, Federal Entities Program, U.S. Census Bureau 10-year July 2019 National Population Estimates; Visualization: CDC CPR DEO Situational Awareness Public Health Service Team

What you're seeing here is historical data from the Centers for Disease Control (CDC), United States Department of Health and Human Services (DHHS). We know these data are historical because just underneath the figure title we see the following date range: August 13, 2021-August 10, 2022. We know these data come from the CDC by observing the blue CDC logo in the upper right-hand corner and also by reading the very tiny, italicized last line of the caption which sites two pieces of information: (1) "Last Updated" and (2) "Data Source". Finally, we know we are looking at data related to COVID-19 vaccine administration within the U.S. by interpretation of the title, "COVID-19 Booster Dose Administrations, United States" (note that we would needed to have known the contextual definitions of booster and dose since the words 'vaccine' and 'vaccination' are not actually used in the title, though they do appear elsewhere within the visualization). We also see that these data have been sorted into age categories (one of many demographic categories): 5-11 yrs., 12-17 yrs., 18-24 yrs., 25-39 yrs., 40-49 yrs., 50-64 yrs., 65-74 yrs. and 75+ yrs. Each age grouping has also been assigned a unique visual indicator created by using solid, dashed or dotted lines in shades of either purple, green or yellow color. This is where visualization of the data begins to take shape. Now we know that the variations of color and intensity of dashes is being used to describe each of the eight uniquely defined age groupings.

The visualization uses two graphs to represent the percentage of COVID-19 booster doses that were administered to people within eight age groupings between August 13, 2021 and August 10, 2022 as reported by the CDC. "Why the need for two graphs in one visualization?", you may ask. The answer becomes apparent when we read the headings above each graph. The heading of the graph on the left reads, "First Booster Doses, % of fully vaccinated people ages 5 years and older". The heading of the graph on the right reads, "Second Booster Doses, % of fully vaccinated people ages 50 years and older". Did you get that?

So to summarize, what we are being presented with visually is a historical look at a comparison of first booster dose administrations (left graph) and second booster dose administrations (right graph) among fully vaccinated Americans broken out into those eight age categories.

I need to pause and briefly describe some basic approaches for analysis of this type of graph. It is very important that we always read the labeling for the "X-axis" and "Y-axis" of the graphs to understand what data are being presented.

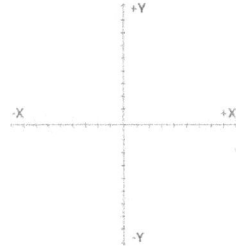

You can find the X-axis by remembering that it is the portion of the graph that is laid out horizontally. Most often, labeling on this portion of the graph will be read left to right. This part of the graph is used to represent a variable that is intentionally changed or chosen.

The Y-axis is the portion of the graph laid out vertically. Labeling on this portion of the graph is usually read vertically starting at the bottom and moving upward. This part of the graph is used to represent a variable that changes in some way based on changes made with the variable chosen for the X-axis. For instance, looking at both our graphs, the X-axis labeling represents the date a COVID-19 booster shot was given ("Date Administered") with a range of August 13, 2021-August 10, 2022 covering the entire X-axis reading left to right. On the other hand, the Y-axis labeling represents the percentage of individuals receiving a booster shot with a range from bottom to top of 0% to 80%. This percentage changes depending on the dates we choose to consider.

Once we've established all of this baseline information (defining the data set), there are any number of ways to proceed with interpreting the data. But believe it or not, you've already done most of the heavy lifting. That's because once you have a fundamental understanding of what information is being presented, it becomes possible to use the power of visualization to see trends and make interpretations and predictions guided by those observed relationships. You're now well on your way to a more fulsome appreciation of both the science and art of data analysis through visualization! Not convinced? See if you can answer the following:

1. As the lines on both graphs move upward regardless of the color, what activity (phenomenon) within our defined data set is increasing? Hint: Look at the label of the Y-axis.

2. As the lines on both graphs move from left to right regardless of the color, what changes? Hint: Read the label of the X-axis.

3. Is there an age range missing? If so, and based on what you've described thus far, how would you identify that age range if it needed to be added to both graphs?

Great job thinking through these questions! You most likely see by the steepness of some of the curves, that different age ranges approached a higher percentage of first booster administration depending on the age range. This makes sense since we recall that the CDC managed the eligibility of both COVID-19 initial vaccinations and boosters by age groupings; starting with older Americans and working backwards in age ranges. But what we can also see from the graph on the left is that as we look at age ranges from older to younger Americans, fewer percentages of each population received a first booster. This is also observed in the graph on the right. Since second booster administrations had only been approved down to age 50 during the time period observed, only three age ranges are present.

In this exercise, we walked through a real-world visualization to become more comfortable with interpreting visual data so we can take note of trends (variations in the physical data) that we might not have otherwise noticed. My strong desire is that you practice interpreting visual data as often as is practical. It's not so important that each bit of data subtleties gets scrutinized and understood, keeping in mind that none of us can know it all. What IS important? Consistent exposure to and practice with the process. Remember, we are not developing this skill simply to assist us in obtaining actionable information from charts, graphs and other visualization tools. Of equal importance is our ability to call upon this same skill set when we practice mindfulness through visualization. The goal: to become more comfortable recognizing and interpreting where we are spiritually and emotionally as we continuously interact with and become shaped by our collective experiences.

Graph on the Left: *First Booster Doses*

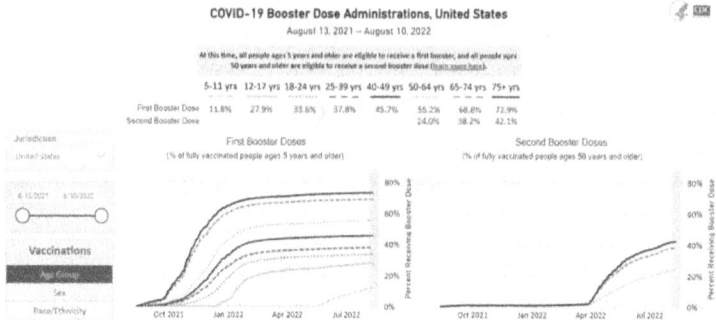

COVID-19 Booster Dose Administrations, United States
August 13, 2021 – August 10, 2022

At this time, all people ages 5 years and older are eligible to receive a first booster, and all people ages 50 years and older are eligible to receive a second booster dose (learn more here).

	5-11 yrs	12-17 yrs	18-24 yrs	25-39 yrs	40-49 yrs	50-64 yrs	65-74 yrs	75+ yrs
First Booster Dose	11.8%	27.9%	33.6%	37.8%	45.7%	55.2%	68.8%	72.9%
Second Booster Dose						24.0%	38.2%	42.1%

The graph on the left represents the percentage First Booster Doses of COVID-19 administered to people 5 years and older. Each color at the top right above where it says "First Booster Dose" represents the age range of people who received the First Booster dosage of COVID-19 vaccination.

Color	Description
	The yellow dashed line shows that **11.8%** of children between ages 5 and 11 received the First Booster dosage of Covid-19 vaccination between April 2022 and July 2022, while no children received a First Booster dosage of Covid-19 vaccination between October 2021 and April 2022.
	The yellow solid line shows that **27.9%** of teenagers between ages 12 and 17 received the First Booster dosage of Covid-19 vaccination between January 2022 and July 2022, while no teenagers received a First Booster dosage of Covid-19 vaccination between October 2021 and January 2022.
	The turquoise dotted line shows that **33.6%** of young adults between ages 18 and 24 received the First Booster dosage of Covid-19 vaccination between October 2021 and July 2022.
	The green dashed line shows that **37.8%** of adults between ages 25 and 39 received the First Booster dosage of Covid-19 vaccination sometime before October 2021 and then between October 2021 and July 2022.
	The green solid line shows that **45.7%** of adults between ages 40 and 49 received the First Booster dosage of Covid-19 vaccination sometime before October 2021 and then between October 2021 and July 2022.
	The dotted lavender line shows that **55.2%** of adults between ages 50 and 64 received the First Booster dosage of Covid-19 vaccination sometime before October 2021 and then between October 2021 and July 2022.
	The dashed purple line shows that **68.8%** of adults between ages 65 and 74 received the First Booster dosage of Covid-19 vaccination sometime before October 2021 and then between October 2021 and July 2022.
	The solid purple line shows that **72.9%** of adults 75 years and older received the First Booster dosage of Covid-19 vaccination sometime before October 2021 and then between October 2021 and July 2022.

Graph on the Right: *Second Booster Doses*

The graph on the right represents the percentage First Booster Doses of COVID-19 administered to people 50 years and older. Each color at the top right above where it says "First Booster

Dose" represents the age range of people who received the First Booster dosage of COVID-19 vaccination.

Color	Description
■ ■ ■ ■ ■ ■ ■	The dotted lavender line shows that **24.0%** of adults between ages 50 and 64 received the Second Booster dosage of Covid-19 vaccination between April 2022 and July 2022, while no adults received a Second Booster dosage of Covid-19 vaccination between October 2021 and April 2022.
▬ ▬ ▬	The dashed purple line shows that **38.2%** of adults between ages 65 and 74 received the Second Booster dosage of Covid-19 vaccination between April 2022 and July 2022, while no adults received a Second Booster dosage of Covid-19 vaccination between October 2021 and April 2022.
▬▬▬▬▬▬	The solid purple line shows that **42.1%** of adults ages 75 years and older received the Second Booster dosage of Covid-19 vaccination between April 2022 and July 2022, while no adults received a Second Booster dosage of Covid-19 vaccination between October 2021 and April 2022.

STEP III. EVALUATE, INTEGRATE, REPLICATE

These last series of steps, evaluate, integrate, and replicate, should ideally be accomplished one after the other in rapid succession. You don't want too much time to pass between each of these three steps. The reason for combining these will become apparent the more you practice. In a nutshell, we are now at the point of taking action. Here is where all of our efforts pay huge dividends. First, we evaluate. We make conclusions based on the data we have just visualized. Second, we integrate. We take action in light of those conclusions. Finally, we replicate. We return to observation mode and repeat.

Feedback Loop

The complete cycle turns into a feedback loop. As scientists, we absolutely love a well-designed feedback loop and here's why: feedback loops are powerful tools to refine and reinforce systems. Once the loop is set, you can run the model continuously. Now, you have a powerful tool of your own to counteract the social media onslaught your children are exposed to daily using some of the same methodology used by the social media giants. For you, as a parent, this means you can use the feedback loop approach to exponentially amplify the positive behaviors you desire from your children!

Remember, you must continuously **evaluate, integrate** then **replicate**. Evaluate whether or not you are getting the desired outcomes with the approaches you are currently using. If the answer is yes, move on to integration. If the answer is no, simply choose a different approach for that situation and with that particular child. Remember, every child is unique. It's ok to not get the approach correct the first time, or the second, or even the third in most cases. Children are very forgiving. They reward loving effort more than anything. They will love you through those early failures, so long as they see your honest effort. Usually, they will end up meeting you halfway. Integration will be fairly intuitive at this point. Simply continue to model those behaviors that are getting positive results, i.e., replicate.

One thing to note here is that you must work to become consistent. Resist the temptation to cut corners on this one, folks. Cutting corners as in not insisting that your child repeat a task that was "half-done", not taking time to answer your child's questions when they are looking for corrective feedback, not offering praise and acknowledgement when your child exhibits behaviors you desire, or by not raising expectations when it is clear your child is ready for more challenging situations. As adults, it may become somewhat annoying, but our children need to see us repeat the same behaviors under similar circumstances. Don't forget that the consistency our children need does not always require a sit-down conversation; nor do we need to always put things in writing (though some children learn better that way). Rather, how we live our lives daily creates some of the most potent examples for them to emulate. They use these nonverbal cues to reinforce those connections to the Truth being hard-wired into them by God, through our parenting. Wow!

Our intentional consistency is the "secret sauce" that God uses to exponentially amplify every Truth we teach our children as they go about the business of **"hiding these words in their hearts, so that they may not sin against God."** (Psalm 119:11) Through the faithful practice of consistency, we help our children to establish their own framework for recognizing, evaluating and responding to Truth. Once our children feel comfortable working from that place of Truth, they will instinctively run to it when engaged in problem-solving throughout their lives. This becomes our enduring gift to our children. Now, let's take a closer look at how practicing consistency can both support and enhance *honest parenting*.

Source

"CDC COVID Data Tracker." *Centers for Disease Control and Prevention*, Centers for Disease Control and Prevention, https://covid.cdc.gov/covid-data-tracker/#vaccination-demographics-trends.

My primary premise in this chapter was this statement:

How in the world are they going to discover that sense of purpose and discover those tools if we are not being honest in our parenting?

a. If you buy that, let me ask you this: When we are NOT being honest, would that mean we are being dishonest? □ Yes □ No

b. Is dishonest parenting really an option in today's world? □ Yes □ No

By now, you must know that *honest parenting* isn't for wimps, chumps, or scary cats! The level of self-reflection alone is enough reason to run and hide, right! Confront my true self? Are you insane? LOL.

a. But, let me ask you: Does the "science" I am teaching you help at all? That is to say, can you see benefits to using a scientific approach as a way to guide you in becoming MORE objective and less emotive when it comes to peeling back layers so you can evolve as an *honest parent*? ☐ Yes ☐ No

b. If so, in what ways and how?

c. If approaching self-reflection using a science-based approach still leaves you "stuck" in defeating thoughts, why/how/in what ways do you believe use of the scientific method fails to guide and inform your experience(s) with self-reflection? ☐ Yes ☐ No

Feel free to send your response to this prompt to dr.buckley@ascientistsapproach.org.

::: Chapter 3 :::
Bring Consistency Into Conflicts

You do not need to be told parenting isn't easy. Commitment to the process can feel like you're in way over your head. This "nervous energy" as I call it, can lead us down two very different paths. I want to focus on the dark path of inconsistency first, because it's important to be able to clearly see signs if you're on this path. Without guardrails actively put in place, it is all too easy to find yourself fighting against the principles you are working so hard to live and parent by. Please don't be fooled.

The dark path of inconsistency has a way of convincing us we are doing just fine, only to later discover we have drifted far from safe harbor.

Yes, you can redirect and get back on track (thank God). But, these fits of start-stop-start-stop can sap the energy out of you and everyone else in your household. Contrary to the dark path of inconsistency, ***honest parenting*** affirms your children's need for a safe, stable, predictable environment in which to grow and develop. They will have many milestones to achieve on their individual roads to physical, emotional and spiritual maturity. What we as parents do to help establish and maintain stability in our children's lives will allow them to pour themselves into becoming who they were created to be without unnecessary distractions and deviations. How do you help create and nurture this consistency?

First of all, recognize and affirm that as a parent, you are not doing this alone. **You are partnering with your children and with God to achieve consistency**. However, stability cannot be forced; in other words, stability is not one-size-fits-all. Stability is personal. It means different things to different people. So, you may not be able to use the same consistency strategies with different children. Secondly, look for opportunities to breathe life and light into your children's everyday lives. As our children grow, life's daily routines can become too routine.

Without realizing, we can easily become complacent and begin to miss vital information our children are providing about the world from their point of view. As their parents, our children's point of view has to remain an important consideration for us, especially as we partner to provide them the appropriate spaces needed to develop and mature. It is not an easy thing to keep your radar up for news from them. But so long as you maintain a desire to do so and practice prayerful meditations, God will provide the answers you need and the tools to use in order to respond in impactful ways.

In the end, that is a major part of the goal of consistency: to approach our children's needs with impactful, loving responses that consistently guide them towards the path God has predestined for them. When we make a conscious, obedient choice to consistently practice *honest parenting*, the end result will be truly satisfying.

Being consistent may be one of parenting's most overlooked and underrated calls to action.

It matters to children that you mean what you say and say what you mean. The world is their playground; their life-sized laboratory if you will. It is your responsibility to help establish and maintain guardrails of consistency for them as they work towards self-sufficiency. Below I provide a scenario that will allow you to observe consistency in action. Let's take a look.

When children meet other children, oftentimes their natural inclination is to make friends, share what they are doing, and begin to play together. It's a beautifully innocent experience to watch. Even more reserved children will be curious about new faces of people who are the same size as they. A typical introduction from a child's perspective would be to say hi and then immediately start asking questions: "What's your name?", "How old are you?", "What are you doing?", "Can I do it with you?", "Will you show me?" or, "Can I try?" They have no problems collecting this data in their unique ways. Human instinct takes over, guiding the flow of conversation and connection such that the children quickly understand each other and set about the business of working and playing together.

Against this backdrop stands the adults. Unfortunately, we tend to insert rules that complicate what would otherwise be a relatively seamless series of introductions, understanding and collaboration. Our first "rule" is usually to insist that the children engage each other cautiously and only with adult permission. Here, we have begun to set the stage for apprehension and fear that would not otherwise be present. The insistence on caution is confusing to a child. "Why on earth do I need to be cautious? This is a little person just like me. I know how to treat them." We are so busy being the "adults in the room," trying to manage social interactions and norms and proclaim them either "successful" or "undesired" that we miss opportunities to let our children explore their world. If, however, we reject those tendencies to control, we will learn how to take advantage of everyday opportunities in mundane situations that push us towards honest parenting. When that happens, **honest parents** become keenly observant guardians into our precious children's galaxies as they discover, again and again, that the things connecting them with other little human beings just like them make them both awesome! Unlike the "adults in the room", their responses of amazement, astonishment, or awe are all natural. The newness and the unknown do not prompt a response of fear or apprehension, but a desire to encounter this human being more fully.

We, as the adults, on the other hand, operating within our worldview, decide to interject undesirable counter productivity (caution, fear, distrust, skepticism), negative emotions rooted not only in our natural survival instincts but also carried over from our own baggage. Now, what once was an opportunity for our children to explore and grow has been reduced to unnecessary conflict in their young, impressionable minds. I can imagine our children asking themselves: "Do I honor my true self and insist on reaching out to this new child (which would mean disobeying my parents)", or "Do I display disobedience to my parents by suppressing my desire to make a new friend?" Why on earth would we demand our children make such a preposterous choice!

Honest parenting insists we don't force a child into unhealthy situations.

Through the practice of consistency, we, as parents, should be constantly evaluating and re-evaluating our thoughts and actions against the standard of Truth.

- **Are my thoughts, my actions, my expectations consistently aligned with the Truth I am working so hard to instill in my child?**

- **Am I living consistently in that Truth?**

- **Am I unafraid to question my own preconceptions and behaviors in order to minimize perpetuating false dichotomies to my children and then asking them to live out MY lies?**

So long as we measure our thoughts and actions against Truth, we will be able to pierce the veil of these manufactured false choices. In doing so, we free our children, and equally important, we free ourselves. When we challenge ourselves to let go of these destructive constructs (although this is certainly not easy, it is necessary), our children usually come up with some amazing responses to their new discoveries!

For parents who desire and choose to travel this road of *honest parenting*, consistency through conflict will require you to adopt a new mindset.

You will have to become intentionally critical about your own ideas and emotions.

Notice I said "become." It's a process. It does not happen overnight or in a vacuum. It is both deliberate and progressive in nature. It requires patience and commitment. Trust and believe me when I say, this...is...not...easy! On many days you will question the process, the investment of time, and energy and your return on investment. You will be tempted to abandon this honest way of life when well-meaning family members and friends question your methods, and sometimes even your motives. Stand firm in those moments.

Know that Truth always, ALWAYS conquers fear.

Fear is the root cause of the apprehension you are experiencing in these moments. Fear that you are not "doing it right"; fear that your child cannot possibly understand, appreciate or even need this *honest parenting* style; fear that somehow you are not truly getting through to your child; fear that you are doing something that will negatively impact your child's development. Yes, as a parent, one of our greatest fears is that we will wake up one day and feel as though we did not give our children what they needed. It's a fear deeply rooted in our own egos as human beings. It's real. But we have to own these fears and then overcome them with Truth. Let me describe a real situation recently related to me that highlights this bedrock principle.

Well, the dog is dead! —
An *Honest Parenting* moment based on a true story...

A family had been dealing with the recent, unexpected illness of their pet dog. This dog had become an integral member of the family; specifically bonding with one of the children. The child and the dog were very close. They shared many important moments in the child's life. The child would whisper secrets only the dog had heard. You could say they were inseparable. As the dog was taken to veterinary clinics for evaluation after evaluation, it became clear to the parents that the probability of death for the family dog was quite high. Of course, this was devastating to the entire family. But it was particularly devastating considering the child had not been exposed to all the ongoing adult conversations with the doctors. Therefore, the child had not been a participant, as the rest of the family was receiving and processing this increasingly dire news about the family dog.

So much "insulating" had occurred (certainly thinking this was the considerate thing to do, and probably from feelings of practicality) that it became more and more emotionally uncomfortable for the "adults in the room" to bring the child into the proper space from which to authentically deal with the current realities. Unwittingly, the parents had created an alternate reality for the child. A reality in which the family dog was simply sick...not near death. An emotionally destructive plan was developed: the parents decided that the best course of action would be to tell the child that the dog was being sent home, with the veterinarian, in order to get better. Now, everyone (except the child) knew that "the dog was being sent home with the veterinarian in order to get better" was adult code for: "We are giving the family dog to the veterinarian to be put to sleep (euthanized) because there is nothing medically that can be done to save him". As the "adults in the room", the temptation to rationalize hiding the dog's eminent demise is that children need to be protected or shielded from death, disappointment, pain and despair. Again, our natural instinct, as parents, is to protect. And so, the adults now have a bigger problem: What happens when the child asks, "How long before our dog is well and can come home?" Who's going to tell the child the family dog is never coming home? When will the child be told? Can you see what these well-meaning, adult rationalizations have produced?

What happens is that as adults, we put ourselves in positions where our trustworthiness can reasonably be called into question...by our own children! Now, guess what the parent's natural response would be? "Look, I am the adult, you are the child. I am not required to explain myself to you! You do not get to hold me accountable for my actions! How dare you question my integrity! My job is to look out for your best interests! I am only doing my job as a parent to protect you from hurt!" None of the feelings associated with these responses are fun; and both the negative responses and hurt feelings could have been avoided by bringing consistency *through honest parenting* to this conflict. Do you clearly see the conflicts here? I ask because there are actually two sides of this conflict;

(1) Variables the adults could not control, or...
(2) Variables the adults could control

On one hand, they had no control over the deterioration of the health of the family dog and what that would mean in terms of the family dynamic. While on the other, they had the choice to be truthful. The truthful path honors **honest parenting.**

Why did I give you the dead dog anecdote? I did so in order to challenge all parents, every one of us, to crave glimpses into the amazement that children and their God-instilled innocence never cease to manufacture. Our children have a resilience and maturity about life that can become muddled by well-meaning adults. More than protecting them from life, we end up preventing them from fully living. What a tragedy! We, as adults, can only wish we could meet situations and conflicts with the bold innocence and courage of most children. I firmly believe that we adults are usually the source that robs children of their God-given abilities to manage all the vagaries of life.

If we remain open and consistent, we will find that God will not only remove unnecessary and controlling tendencies from us, but He will also remind and return us to the safety of that innocent child living within each of us. We must become open to the idea of protecting them against our "adultness" and the behaviors we have adopted and convinced ourselves are "appropriate" and "safe."

The goal is MORE self-disclosure from our children, not LESS. The goal is to look for and embrace opportunities to be consistent in relating to our children. Our goal is honest revelation, not retreat.

Practice, practice, practice. Together, honesty and consistency will produce the desired outcome. You have to trust that. But, before I close this chapter, we need to talk about what may be the absolutely toughest conflict for us to admit let alone confront, i.e. the conflicts we, as parents, create in the process of *honest parenting* because of what I call "-isms." I want to take my time and explain.

Your Personal Ism's:
Instinctively Self-serving Mannerisms

There are continuous and revelatory layers within the framework of **honest parenting**. Which means, over time, we discover more and more truths about ourselves and our children. Those truths and those outcomes are literally "baked" into the **honest parenting** process. Here's why:

In order to truly parent honestly, you have to:

(1) Be brutally honest with yourself as you become aware of ideas and beliefs (your "-isms") you've held that just don't ring true or make sense in the context of love and compassion for humanity.

(2) Be vigilant about safeguarding your children from your natural human tendencies to project (intentionally or unintentionally) these "-isms" onto them...even as you continue the challenges of working through them.

(3) Be able to celebrate (or at the very least accept) the reality of your "isms".

(4) Be aware that this whole process is a life-long journey; one that you and your children will find of benefit...so long as you stay the course, respect the process and listen to both God and your children.

As I said previously, when we slow down just a bit (just a tiny bit, ya'll) and listen to our children patiently, it is remarkable how readily they will provide the information and direction we, as parents, need in order to properly navigate them through their early developmental years. It is our responsibility to recognize and affirm - to ourselves and to our children - that the formation of their value systems should not simply be reduced to adherence to some "parental mandate" handed down by us as the all-knowing adults. Rather, honest parenting insists that we partner with our children as they set about the exploration of this wonderfully, richly diverse and yes, sometimes confusing world. We have a fundamental choice to make as parents:

Is our goal to raise children who view the world the way we do because we forced our "-isms" on them? Or, is our goal to raise children who can truly see the world for themselves, think critically about the realities they see, use their own reduction method so that they can make independent, informed assessments and decisions that honor the values they have chosen to adopt?

Yes! I want to tell every parent that there is nothing wrong with explaining to your children your views and why you hold them. But there is absolutely something wrong with your deciding role, as a parent, to insist that your child believes as you believe, without question, and with no desire to allow them to come to conclusions for themselves. Even God refuses to force Truth upon us. Instead, we have been given freewill. Why would we not lovingly offer that same freewill to our children...and teach them to always, always respect this most precious gift of God that embodies and exemplifies pure, honest, unbounded love (Deuteronomy 30:15-20; Joshua 24:15).

Finally, I want you to know that I know this is just a book, simply words on a page; and, that despite my convictions that because *honest parenting* makes sense and works, I am in no way saying that I am anyone's live-in-nanny, personal parenting coach, or some know-it-all family therapist available to meet with you on a schedule in order to condition or train you for the rigors of your *honest parenting* journey. But I am genuine in my hope that each of you will want to focus, at least once a week, on making an effort to get out of the way of your child's playtime...especially playtime with other children. I understand adult supervision is a must. That's not an issue for me, nor for the children. They welcome adult supervision. They have an instinctive expectation for appropriate adult intervention. But other than that, let your child be a child. Other children are very capable of setting boundaries and engaging in compromise with minimal adult intervention. It is certainly not necessary for us, as parents, to constantly decide, for our children, how they should relate to other children.

Our job should be practicing *honest parenting*, properly modeling for them how to behave and treat other human beings.

They can then model those behaviors in their own world, among their own peer group.

Yes, your children will encounter other children who have not been exposed to the same models of behavioral norms. This will present a definite challenge because this is typically the origins of bullying behaviors. Since a child's willingness to bully is rooted in modeled adult behaviors, it becomes a tricky proposition to help a child navigate these situations.

You cannot simply ask your child to ignore the bully.
They already know that doesn't work.

Plus, they naturally don't want to ignore another child they've decided to engage. They want to make a new friend and begin to learn from them. It's all about new experiences and learning. So, they are quite curious about this new behavior. This is why you will see children continue to make attempts to associate with a child that they KNOW has been bullying them. They are trying to learn about this child. Our children are simply trying to "crack the code" and learn to speak this other child's language.

Two additional challenges for children to master are sharing and caring. It's tough for children to deal with these because for about the first three years of life, our responsibility to provide for them causes us to paint a very unbalanced picture of sharing and caring. Our children are programmed, by us, to assume that everything is theirs. Think about it. Bottles: theirs; diapers: theirs; toys: theirs; your cellphone: theirs; your attention: theirs. Even when there is more than one child in the home, individual ownership is instilled early and often. No wonder children are confused and angry when told they must share or wait their turn! But mastering sharing and caring is an important stage of development that will go a long way in determining whether or not your child will opt for bullying as a way to settle conflicts or employ the conflict resolution method discussed in chapter 2. It's our job as **honest parents** to present them with alternatives (sharing and caring), making those options the more **HONEST** choices.

A child is being raised by his/her same-sex parents (male or female...doesn't matter). The child has become old enough to begin to notice differences between their family and those of their friends and playmates (other kids referring to "mommy and daddy" when speaking about their parents, Mother's Day and Father's Day cards not really making sense to them, not knowing how to respond when a teacher or caregiver asks if Daddy is going to come to PTA meeting, etc.).

a. What should the parents look out for when trying to determine how much detail to give their child, relative to the US social context of levels of acceptance of homosexuality?

b. What, if any, duty do the parents have to engage other adults in "positions of trust"; explaining how they will raise their child, and helping the adults know what language and approaches are considered appropriate for their family?

This same child befriends several of her classmates. During the school year, her parents make plans for her birthday party. One of the child's best friends has parents who are evangelical Christian. Their denominational belief system is both clear and sincere: homosexuality is a sin; not according to them, but according to God. God will punish such sin by eternal damnation.

    a. What are the pros and cons of sending a birthday invitation to their daughter's best friend?

    b. Do the evangelical Christian parents have an obligation to explain homosexuality to their child? □ Yes □ No

    c. Whether or not the child attends her best friend's birthday party, do the parents involved have an obligation to begin a frank discussion with one another since their children are best friends?

::: Chapter 4 :::
An Honest Approach to Learning
by Micah Murdock

Support Your Child's Learning

In chapter 3, we discussed ways to use consistency to help address the challenges of parenting through conflict. Another critical element of Honest Parenting is to participate in your child's learning as fully as your circumstances allow. Regardless of whether your child attends a public school, private school, charter school or is homeschooled, as a parent, you play a critical role in the success of your child's learning and their overall life preparation. Let's look at a few simple ideas of how to support your child in their learning from infancy until they leave home.

Foster an Attitude of Curiosity

The drive to pursue learning goals is strongly influenced by the level of intrinsic motivation of the learner. It has been said that no amount of skill, coercion or persuasion can force an unwilling student to learn. On the other hand, no amount of difficulty, obstacle or challenge can keep a willing student from learning. Curiosity is an essential characteristic to learning as it is the primary fuel of our intrinsic desire to learn.

In our day, quick and easy access to practically any topic imaginable, gives a curious (or intrinsically motivated) learner, all the opportunity they need to learn a great deal about whatever topic they are interested in. Some examples might include: learning to play an instrument, building a robot, improving athletic skills, repairing a car engine, baking a cake, calculating the area of a triangle, discovering the cause of an international conflict, learning about a country's history, becoming proficient in another language, writing a software program, editing a photo, designing a rocket, donating to a cause and many, many more. With digital information access becoming more ubiquitous ("digital divides" aside), the question is becoming less about "how to get information into the hands of our children," but instead, "how to teach them effective ways to apply it in the real world." This focus on application will be addressed later in the chapter.

First, let's talk about how to foster the curiosity of our children. The good news is children are born naturally curious and unless trained otherwise, will remain naturally curious throughout their lives. The bad news is, that to increase the efficiency of learning in our classrooms, we can, and often do, unintentionally replace their natural born inclinations toward curiosity with expectations of compliance and standardization. Our school systems could unwittingly be a significant contributor in reinforcing this "curiosity-for-compliance" trade off, as they are often guilty of overscheduling and over-structuring the learning process. Excessive focus on grades, GPAs and standardized exam scores, for purposes of assessment and progress reporting, may serve to stifle the basic creativity and curiosity of our children. This must not be allowed to happen since it is our children who provide the uniquely creative catalysts that ultimately drive the advancement of society.

We can guard against this at home by providing regular periods of unstructured learning and by encouraging our children to ask questions and search for answers on their own. We should also challenge our children when we hear them making statements or observations that are overly simplistic, too broadly applied, or do not reflect an acknowledgement of the myriad, diverse views held by others also living and working in the world in which we actually live.

For instance, when I hear my children parroting things they've heard from friends, the Internet or even teachers at school that could be seen as inaccurate, biased, or too simplistic, I challenge them by asking questions, and then encourage them to participate in finding more complete information. This simple and often brief exchange encourages my children to question what they don't understand, and then teaches them how to find answers on their own. I hear my 8-year-old regularly saying "Alexa®, what is…" fill in the blank. While Alexa® doesn't always have the best answer to his questions, it's a demonstration of curiosity and develops habits that will hopefully remain with him throughout his life.

Be Present and Set Healthy Expectations

Being present in our children's lives does not mean we are with them all the time. Most children would not benefit from this type of parenting anyway, especially as they get older. They must be given room to make their own decisions and learn from the mistakes they make. What it does mean is that we are involved in their lives enough to have a working perspective of what is important to them, and then regularly engage them through this perspective. Of course, this includes school, but it also includes extracurricular activities and the events of their social life. If your children are interested in something that isn't important to you, you could exemplify the principle of curiosity and spend some time learning more about those things that are of great interest to them.

As an example, I grew up in an athletic family. Most of what my family did outside of school revolved almost exclusively around practices, games, and tournaments. So, when one of my children expressed an interest in theater, I had no prior personal context. Instead of trying to coerce her into trying harder during her soccer practices (which she clearly hated), I asked her to teach me about the world of theater. She felt heard by seeing that I valued what was important to her, and she felt appreciated when she had a chance to teach me something I didn't know anything about. My regular, but brief inquiries into something that was important to her, even though I had no interest in it myself, helped me be present in my daughter's life. As a bonus, I've also found that through my daughter's willingness to expose me to the world of theater, I have developed a new personal interest that I very much enjoy.

Here are some other suggestions that can help support ways to be present and set healthy expectations for your children.

Establish Routines

All humans operate more effectively when there are clear expectations that are consistently enforced. Establishing before and after-school routines for how your child is expected to get ready for their day and complete their homework, reduces the need to repeatedly describe expectations and increases the chances that your child will respond in a productive way. These routines serve as a foundation for many of the remaining principles suggested in this chapter. When your child has a regular routine and can predict what to expect regarding their daily activities or potential results of their behavior, there is more trust and better opportunities for real communication.

Ask Specific Questions About Your Child's Day at School

No one questions the importance of talking with your children on a regular basis as a means of deepening your relationship with them. This was emphasized in the section on being present. However, if your efforts to start a conversation rely primarily on the commonly asked question, "How was your day?", it's likely that you have discovered how hard it is to get your child to say anything more than "fine", "good" or the classic shoulder shrug.

While this is a good attempt at being involved in your child's life, it rarely produces the types of responses that generate a conversation.

Asking very specific questions such as, "How did you do on your math test", or "What was the best thing that happened today?" or "What did you do in gym class?", can lead to more detailed responses and can lead to more meaningful discussions. Of course, when you ask questions like this, you need to have a previous knowledge that those questions are relevant. If you don't know your child had a math test today, it's hard to ask the question. The more consistently you ask detailed questions, the more you will discover, and the more detailed your questions can be the next day. This can snowball into effortless conversations and can even lead to your child volunteering information without you having to pry. As was said earlier, open, ongoing communication is essential to being present in your child's life.

Demand Effort, But Don't Punish Your Child for Getting the Wrong Answer

Many of our schools are better designed to teach our child how to pass an exam than they are to prepare them for life after school. Considering the rapid pace at which our workforce evolves (post-COVID trade and commerce being a prime example), specific facts and information our children learn in the classroom are becoming outdated at a similarly rapid pace. The skills that are becoming more important are those that develop the child's ability to adapt to this ever-changing world. We should be pushing specific facts and figures less, and pushing good fact-finding skills and strategies more. We should focus less on whether the learner can regurgitate what was on their test review, and more on whether they can identify and solve relevant problems with the information they have at their fingertips. Conversely, we should care less about their ability to comply with outdated classroom policies that will not serve them in the workforce or in their personal lives, and more about their ability to effectively communicate and collaborate to solve the complex and relevant problems of our day.

"But getting good grades is important to my child getting into the best colleges", you may say. Two things to remember about that statement is that, (1) grades are only one measurement of achievement, and (2) successfully completing college is only one avenue, of many, designed to assist in laying the groundwork towards a successful career and life. There are enough uber successful people in this world without a college degree to put a serious question mark on the assumption that an expensive and often outdated college degree is still the only way to make it in the world. Does this mean we should not care about our child's performance in school? Of course not. However, it does give a good argument to focus less on grades and more on effort. Chances are, if your child is giving their very best effort, they will get good grades. As their focus shifts more heavily towards the quantity and quality of the effort they put into the learning enterprise, rather than prescribed outcomes, when they do occasionally bomb a test or fail a project, they will know it isn't the end of all their hopes and dreams. Although their GPA may dip slightly, they will still own all the benefits that come with having a strong work ethic and real-world experiences that encourage them to try again; just like they will in practically every job they will ever have.

Help Your Child Develop the
"Three Characteristics of Lifelong Achievement"

This last point brings us to a deeper discussion of the question, "What really does predict and reliably define our child's ability to find success in life?" Much research has gone into the understanding of achievement and high performance. Among the conversations on the topic, three overlapping characteristics surface repeatedly.

They are:

Let's talk about each of them individually.

Grit

Grit is defined by Angel Duckworth as,

"Passion and perseverance for long term goals." [1]

Duckworth goes on to describe what grit is not.

"Grit isn't talent. Grit isn't luck. Grit isn't how intensely, for the moment, you want something." [2]

With this understanding of grit, we better understand how helping our children develop grit, as a primary characteristic, would have a positive impact on their lifelong success. For example, a gritty child who gets only average grades, but can overcome a learning obstacle, or a difficult personal circumstance is more likely to achieve greater lifelong success than a child with no grit but who achieves high marks because they are gifted at memorizing the answers to an exam the night before. The benefits of learning early in life that grit can help your child overcome any challenge is infinitely more valuable than learning that they can pass a school exam with flying colors even though they waited until the last minute and crammed their way though.

Discipline

Discipline is a close cousin to grit but emphasizes an important distinction. While grit is all about passion and perseverance, discipline focuses on the importance of consistency. Adding steady effort (discipline) together with passionate perseverance (grit), will demonstrate to your child that, these two characteristics can help them accomplish whatever goals they focus on. Supporting your child in establishing worthwhile goals, and then utilizing the reduction method or helping them break those goals down into simple daily habits will, over time, create success in whatever their chosen area, and more importantly, will help them gain confidence in themselves and their ability to achieve and impact the world around them. In essence, discipline is manifested by us partnering with our children in their daily habits that, in turn, work in concert to create the life we build for our children over time. The more disciplined or intentional we are about our own habits while simultaneously helping our children establish their own habits, the more likely our children will end up achieving the goals that we both set for their young lives.

Growth Mindset

This leads us to our last characteristic, Growth Mindset. Carol Dweck, the author of *Mindset: the New Psychology of Success*, defines Growth Mindset as,

66

"people [who] believe that their most basic abilities can be developed through dedication and hard work..."

Or in other words, our abilities are not fixed at birth. This also implies that our abilities are not limited by those things that seem to come easy or "naturally" to us. All talent and abilities can be further developed through effort over time, through grit and discipline. Embracing a growth mindset supplies the mental conviction that through grit and discipline I become better at something I am not good at yet. Without a growth mindset, the fundamental principles of grit and discipline fall apart. Any one of these can exist independently, but the combination of these three critical characteristics over time can provide a pathway, for our children, to success for practically any endeavor.

In summary, some fundamental principles of using Honest Parenting to help our children become more effective learners include (1) fostering creativity, (2) being present and setting healthy expectations, and (3) promoting the development of the three characteristics of lifelong achievement: grit, discipline, and a growth mindset.

Source

1, 2 - Duckworth, Angela. *Grit: The Power of Passion and Perseverance*. Paula Wiseman Books, 2020.

::: Chapter 5 :::
Check Your Tech and Other Parental Safeguards

Honestly Parenting Alongside Today's Technology

Our millennial children are fearless when it comes to technology. It's actually one of the character traits that define the generation. While we, as parents, are debating the merits of wireless and Bluetooth® devices, smart appliances, and connected homes, our millennial babies are forging full steam ahead with kid-created YouTube® channels, augmented reality gaming, and virtually connected lives (think FaceTime®, Skype®, Google Duo® and Instagram®, to name a few). The pace and level of integration of tech will continue to grow. What might this mean for those of us striving to utilize honest parenting approaches to create safe spaces for our children to grow and develop? First let's take a quick look at what tech you can expect them to be living with now and in the coming years. Then, we will explore how to control the influence of technology on your *honest parenting* efforts.

Millennials lead on some technology adoption measures, but Boomers and Gen Xers are also heavy adopters

% of U.S. adults in each generation who say they ...

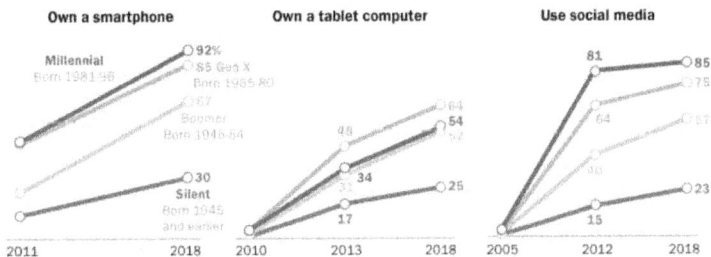

Source: Survey conducted Jan. 3-10, 2018. Trend data are from previous Pew Research Center surveys.

PEW RESEARCH CENTER

Our smartphones are our lifelines. Many homes now do not keep a traditional landline for telephone service, opting instead for cellular service plans for the entire family. Smartphones and tablets are increasingly being used as the primary choice for consuming media content: news, weather, sports, movies, videos...all at our fingertips. You'd have to think long and hard to remember exactly how to get photos from photographic film (if you even still use traditional film). We've become comfortable taking and sharing photos and videos of just about every imaginable aspect of our lives and the lives of our children. What used to be considered our private lives, we now regularly and freely post as public consumer content for the world to see and provide commentary.

The digital age allows us to simply speak to set alarms, answer the door, pick a movie, listen to a song, obtain directions, make reservations, hail an Uber®, order dinner...the list is becoming endless. As convenience becomes our society's mantra, we are more easily choosing to look the other way as technology begins to be used by us as a crutch. We begin to convince ourselves that propping up a toddler with a smartphone is ok...so long as the content is "educational". We convince ourselves that we had a wonderful family dinner when in reality everyone was focused on their smart devices for most of the time. We figure it's one less thing we have to discuss when we allow Google Assistant® to define and spell words for our children instead of them asking mom or dad. We retreat into our bedrooms and leave the kids with their smart tablets and their active internet accounts. But hey, we limit the screen-time so it's ok, right?

These are tough situations and there really is no one answer that covers all. Each of you will have to chart this course on your own and make tech work with and for you and your children, not against you. But I can offer three (3) truths for you to adopt.

TRUTH 1

You're not going to rid your children of technology and you shouldn't want to do so.

If something offends us, the easy response is to get rid of it. It's less time consuming than trying to figure out the root cause of the offense. We feel safe in believing, "out of sight, out of mind." But we know inside that this mantra is far from true. Current and future technology forces us to search for a new approach because it's a losing proposition to think we can simply throw out the offending technology. Go ahead. Let me see you try that one!

In today's world, the integration of man and technology forges ahead at breakneck speed. We marvel in awe at the latest gadgets and dutifully set about the business of incorporating them into our lives. Seriously, how many of you reading this book are still amazed that at the time of this book's publication, you can purchase refrigerators that will alert you by cellphone when food items are low AND that you can purchase cars that park themselves and, in some cases, drive themselves! Any parent who is fighting technology is fighting a losing battle. Let me add an additional consideration: Limiting children's exposure to technology will surely put them at a disadvantage as they grow into adulthood. As *honest parents*, that is the last thing we want for our children. A more useful approach is to model and set expectations for responsible use of technology versus creating unnecessary conflict that risks alienating your child.

Technology alone will never be a viable substitute for authentic human interactions.

I have witnessed many bonding sessions in which a child explains technology to their parent(s). My favorite example of this type of parent-child bonding is when a parent has to ask their child for assistance with a new piece of technology hardware (e.g., new model smartphone, new smart home device, new healthcare/fitness device) or a software application. If the parent postures themselves as a learner, there is a rich connection with their child, the technology lesson facilitates a conversation that initially may have felt more like an uncomfortable crutch. Pushing aside all status, pride and ego concerns the "learner" (i.e. parent) asks the technology facilitator (i.e. the child):

- "So, how did you learn about this?"
- "What do you like/dislike about it?"
- "How do you think this tech could be misused?"
- "What sort of privacy issues might be associated with this tech and how should we manage this as a family?"

You get where I am going with this, right? The conversation-starters during this type of parent-child bonding are potentially endless. The possibilities suggest that when we give children the choice between positive, human interactions and their access to technology, the child will choose positive, human interactions virtually every time!

Honest parents believe that technology will never be able to fully replace our children's need of us for clarity, insight, affirmation or motivational needs which only human interaction can provide.

Once all of us, as parents, understand this, we will be able to responsibly manage technology use within our family as a companion of **honest parenting**, but never a substitute.

✦ Put a pin right there, please. I want to go into a "deeper dive" into this second truth right here. I am going to apply this second truth to a topic we covered in chapter 3, i.e. bullying. In so doing, I hope to really underscore the importance of behavior modeling which at the core is the entire essence of our second truth regarding honest parenting's insistence on AUTHENTIC human interaction. My bullying application of this second truth goes like this...

In chapter three on consistency, I tackled bullying behaviors among children and the need for us, as parents, to be able to identify the early, warning signs of bullying. I brought up the topic because I feel it is important to alert my readers that not only should we have an intolerance for our own children being bullied, but we should have an equal disdain for our children becoming the neighborhood bully! From that space, I offered you **honest parenting** strategies, along with my in-your-face challenge—i.e., to model the positive behaviors you ask your children to display. Now, after letting that idea marinate, I invite you into what I call the "deeper dive" by asking you to consider another case. How would an honest parent even begin to peel back the layers if their precious angel had, in fact, been bullying someone else's child?

I have some thoughts about how an otherwise good-natured, enjoyable, and unintentional child might have learned how to bully other children. I want to ask you a question: **Are you "bullying" your children?** Be honest. Are you constantly demanding of your child what they must do, leaving no room but for them to follow expectations, rules, directions, and norms that you have established, modeled, and reinforced? If force and threat, i.e. parental "power" is your customary go-to, then the answer is probably, yes. You are "bullying" your own child. They are learning *from you* to assert their desire and sway people to compliance based on power dynamics versus earned trust, consistency, and fidelity. In other words, you relate to your child as an authority with power to wield consequences much more than a faithful person who has earns and deserves their trust as well as their obedience.

In sum, your position of power, as the parent, communicates the message that they HAVE to agree with you, your opinions or your way of approaching life. Their compliance results in approval, acceptance, and continuation of the connection—a connection that should be a relationship, but functions more so like a concession. Likewise in their connections (not true relationships) with their peers, they figure out in their own little brains that the one who complies (i.e. the weaker one who seeks approval and acceptance), must or has to agree with the one who behaves as if they are in control (i.e. the strongest one).

I want to be clear. I am not suggesting that children and parents are in an equitable relationship in which we must negotiate terms. Rather, that when we recognize ourselves as stewards, and as our children's first teachers, that does not mean we have an open-ended right to impose our will on them with absolute power and no regard for how the exercise of our parental power communicates lessons about how to treat other people. I want to help us through the difficult and delicate and potentially destructive balancing act of knowing how to use parent power honestly. Perhaps my goal would be best served with a case in point.

At some point in their development between pre-school and early adolescence, all of our children ultimately encounter children (and sadly sometimes also adults), who just don't "play by the rules." Those defiant, discourteous, and divergent children usually have picked up their destructive behaviors from somewhere, i.e. from within their own family, their friends, their caregivers, digital media or online content, etc. which entices them to act in ways that are deceitful and disrespectful. Such children are intentionally bent on breaking rules, making other kids feel miserable, and are emotionally (as well as sometimes physically) defeating other children. What do you do when your child comes home from school and relate that first story to you that clearly indicates some other child (or group of children) at school has decided to be untruthful, disrespectful, or teasing at your child's expense? This is actually a tough situation. We know we cannot force those bullies to comply with the rules for proper treatment of other human beings; we can pray for them, but that's about it. Similarly, we know we should not allow children to bully nor should we turn a blind eye to the parents who support this type of behavior.

How do we navigate such a quandary, both for ourselves and our children? How do we embrace and actualize honest parenting principles even in the face of adults and children who break the rules? Well, first start at home. Recognize that your sphere of influence may not readily encompass the rule-breakers. But, you most certainly have the ability to influence your child. Start there. Remind your child that above all, they are loved. Work from that position of strength. Next, listen. Listen to your child. Give them more than one opportunity to describe what they are experiencing. Take time with your child. Seriously. Sometimes it takes a few tries for children to express themselves fully. Don't rush them, or you may not get the full picture. Now, take off your logical, "I've gotta fix it", parenting hat for a bit. For now, simply try to experience what your child is describing to you and I guarantee you this: God will take you, your child and those experiences, and weave together a beautifully crafted tapestry of love and respect that you can then use to teach your child how to consistently face this type of conflict with the confidence in their worth and your love.

Experience their hurt, their embarrassment, their anger, their confusion. Take it all in, without reacting to all of those emotions. Just experience it fully, without judgment. So, while the temptation is to exercise parental power in absolute ways to protect our children and ensure their safety, the harder yet more substantive relationship-building technique, rooted in trust, could become the source from which you craft the most appropriate responses to conflicts in your child's world. I honestly believe your child will appreciate and respect you for remaining consistent, rewarding you not only with their trust and affection, but also obedience in the process and your practice of honest parenting—even when they find themselves surrounded by other children who are bullies or rule-breakers. Please remember that:

**"Power tends to corrupt;
absolute power corrupts absolutely."**
- Lord Acton, British Historian

Technology must be a safe space for our children's healthy development.

It is our responsibility as **honest parents** to provide our children the safety they need as they engage technology. I have argued, throughout this book, that children are naturally curious and naturally unafraid, i.e., explorers. These innocent qualities that allow our children to be as open as possible to the world around them in order that they may maximize their development, are the same qualities that can cause them harm when exploited by people who have lost their moral compass. This condition can be applied to just about all technology associated primarily or tangentially with personal data acquisition; which covers a lot of technology sectors. For now, though, I want to touch on just one of these: the Internet and associated social media platforms.

This is probably the toughest portion of the book for me. It's tough for me to write, for the same reasons it's likely tough for you to read. We have to have a frank discussion about the fact, that as parents, we are scared to death of these psychopaths, who we know are lurking online to prey on our precious children. It's a dangerously wicked threat to all of our communities. What's worse is that we as parents feel virtually helpless to combat the problem. I struggle with you on this issue constantly. We cannot turn back the clock and stop the rise of social media. That train has long since left the station. We see story after heart-breaking, gut wrenching, sometimes outright sickening story of children being redirected by predators from innocent, age-appropriate material to inappropriate exploitative content; or teenagers struggling with identity theft and cyberbullying on social media platforms, or our babies being deceived into believing they are making a new friend, only to be abducted at the hands of pure evil! I could go on, but you get the picture...we all abhor these monsters. Yet, we know the same technology we encourage our children to embrace and own also is home to some of the worst versions of humanity imaginable. What right-minded parent wouldn't protect their children from that evil?

Taken a step further, we fall into this trap: "Any parent who DOESN'T take the necessary steps to protect their children is simply irresponsible and a disgrace to all of us good, honest parents." Really? Do you really want to pass this type of judgment? Of course not, because we know the issue is much more complex. The truth is that the cost of implementing the needed levels of protection to prevent these unthinkable harms to all internet and social media consumers is well beyond the financial, technical and educational levels of the majority of consumers of the technology.

Look, I have no desire to become an expert in cybersecurity. I do not want a second job as an information technology systems or security specialist. For most of you reading, my guess is neither do you. And yet, this seems to be what Silicon Valley is selling us, as parents and consumers, in this digital age. Tech companies who make billions, annually, through essentially buying, selling, storing and analyzing OUR personal data have the audacity to tell us that WE must bear the costs associated with keeping that data, and our children, safe and protected. So, we freely give them the data they are constantly in need of to drive their business model and in "exchange", we then get to pay them to protect that information. Oh, and by the way, those same tech giants take it a step further by informing us that it's OUR responsibility to manage all of the attendant IT security issues inherent in the deployment and integration of all this technology; they say it's our financial AND intellectual responsibility.

To add insult to injury, we as parents are led like sheep to slaughter, believing the lie that we are bad parents if the whole exercise becomes such an overwhelming proposition that we simply throw our hands up in frustration. They've really done a number on us since they know full well that the vast majority of us have zero knowledge of how to properly secure and maintain the digital networks over which all of this information flows. News Flash: we as consumers should not be held responsible for policing the Internet or securing the myriad of technologies, we have to use in order to function as productive citizens. Furthermore, we as consumers and as parents of children who themselves are consumers, should expect (and demand) that those who make billions from our willingness and economic ability to access technology be REQUIRED to spend the money necessary to secure and protect the entire system. To me, it boils done to this: If your business success requires access to the personal data of we the consumers (and those who aren't even primary consumers!), and you get that data FREE, then the very LEAST you can do is pay to protect the integrity of the system. Full Stop.

I cannot leave this discussion of parenting and technology without addressing the issue of the growing "digital divide" in this country which mirrors the "global digital divide". Technology is so pervasive not only in the U.S., but in every developed nation such that children and their families who lack appropriate access are quickly becoming part of the new "Technology Ghettos". It is a sad commentary on our collective humanity and the power-hungry tug of capitalism that this latest iteration of classism is being allowed to take root.

The American Recovery and Reinvestment Act enacted by congress and signed into law by President Barack Obama in 2009 allocated billions of dollars to government agencies in an attempt to significantly increase broadband service coverage across the United States, especially in rural and low-income communities. The results were marginally successful. Yet the gap between the information and communication technology haves ("ICT haves") versus have-nots ("ICT have-nots") has continued to grow; so much so, that the language of the conversation now focuses on "digital inclusion" rather than the divide itself. Global technology giants clearly have a role to play in solving this problem. But tech companies should not shoulder this burden alone. I recognize this is not one issue, but a complex series of interconnected issues. Further, when questions surrounding who will bear the costs associated with digital inclusion and how best to deploy, upgrade and integrate the underlying infrastructure are added, then the complexity of these issues increases dramatically. Still, we cannot afford to ignore these problems by declaring "It's too complex of an issue to solve". A full embrace of honest parenting leads to the real and obvious truth: for the sake of our children, we as citizens MUST make the conscious decision that access to information and communications technology should be a universal right of ALL. Then we must turn that truth into action. Our children will eventually see that we truly, honestly live by and through the timeless TRUTH: **"Faith without works is dead."**

Chapter 5: Deeper Dive Discussion

Prompt 1

In this chapter, you read that honest parents should not need a Ph.D. in cybersecurity to efficiently and effectively protect our children from online predators as well as other security concerns.

a. If you agree that our children's safety is a right, not a privilege, what collective strategies should we, as honest parents, pursue to deliver these protections to all children?

b. In addition to our due diligence individually within our respective families, what more can we do as a community?

As members of the *Honest Parenting* community, we know that truly listening to our children in a patient, loving posture allows them the freedom they need to realize who they are and then relay that information to us, so that we can continue to be effective parents (another "feedback loop!"). Similarly, our children's insights could inform these technology executives and private venture capital firms that finance the former.

    a. If you agree that our children's safety is a right, not a privilege, what collective strategies should we, as honest parents, pursue to deliver these protections to all children?

    b. In addition to our due diligence individually within our respective families, what more can we do as a community?

::: Chapter 6 :::
Parenting Through COVID-19

A s of this writing (winter/spring 2022), Coronavirus Infectious Disease has claimed over 900,000 U.S. lives and over 5,500,000 lives around the globe. It has become the deadliest infectious outbreak world-wide since the Influenza pandemic of 1918. SARS-CoV-2, the official name for the coronavirus responsible for this global pandemic, was first characterized in a cohort of illnesses documented in Wuhan, China (December 2019). Since then, our modern world has been ravaged by devastating upper respiratory symptoms ranging from those related to cold/flu, to novel loss of taste/smell, to life-threatening pneumonia. The sheer numbers of those infected has wreaked untold havoc on healthcare systems worldwide. Many governments raced to severely restrict movements of citizens, in their respective countries, in an effort to slow the pace of rising infections that, if allowed to run through populations unabated, could easily cripple even the most sophisticated healthcare facilities and the critical care staff on which they stand.

We, as a global community, have witnessed first-hand the varied approaches employed by governments in response to this public health crisis. These policies span from one extreme which include minimal government restrictions designed to allow the virus to naturally infect until so-called "herd immunity" is achieved (e.g., Iranian and Swedish governments), to the other extreme: large-scale lockdowns of citizens designed to restrict movement, thereby reducing opportunities for the virus to infect (e.g., U.S., British and Chinese governments).

Two years into this pandemic, and we now have multiple vaccines developed and available to varying degrees within countries. We have largely overcome the early shortages of personal protective equipment (PPE) such as high filtration masks, disposable gloves and hand sanitizer. Two pharmaceutical companies have even developed antiviral pills that have been shown to significantly lessen the severity of illness should an individual becomes infected. All these achievements, especially the speed of vaccine development and deployment, represent successes. And yet, as a global society, we still find ourselves in large part living at the whims of a virus that shows no immediate signs of going extinct.

New viral variants, naturally occurring mutated versions of the original SARS-CoV-2 possessing enhanced combinations of infectivity and virulence, continue to arise sending communities back into various levels of lockdowns or exposing them to constant "waves of new infection". We have become almost oblivious to the daily case counts of new viral infections, inevitable hospitalization rate increases and accumulating death tolls. The collective human psyche cries out, "ENOUGH!!!" Seemingly caught in the middle of this mayhem - our children.

In the early days, weeks and months of the pandemic, we all watched with a collective sigh of relief as the accumulating scientific evidence suggested our children would likely be spared significant illness and/or death from the infection. Then came "the variants". One by one, we were introduced to Alpha, Beta, Gamma and most recently, the Delta variant of the coronavirus. With each variant came ominous cautionary reports: increased infectiousness, increased virulence, more complex symptoms and nuanced complications. Our once comforted minds were once again thrown into panic overdrive. "Shutdown the schools!", "Mandate vaccination for all those interacting with our children!", "Order universal indoor masking in schools!", and "Where are the vaccines for our children!?"

Call after call came to protect our children; sometimes with conflicting demands. "Mask up!", some implored. "Free our children from suffocating masks!", demanded others. "Keep our children in remote learning environments!", was one cry. "Our children are suffering irreparable damage from being deprived of the social/emotional interactions that come with in-person learning!", warned others. Day after day after day we were inundated with new studies, new findings, new recommendations and new worries.

As parents, we have worked to sift through all of this oftentimes seemingly contradictory information, trying to ascertain and implement what we believed to be the best decisions for our families. We have worried late into the nights and early into the mornings, praying that we get it right. We have been told daily that our very lives and those of our children depend, in no small part, on the decisions we make–related to a virus we never wished for, which apparently originated (at least in part), from a city in China that many of us had never even heard of before December 2019. How in the world could we "get it right" under such pressure?

Fast-forward to now, winter/spring 2022, over two years later - and we HAVE survived! We DID (at least in part) get it right. And God IS good! While none of us would be bold enough to claim to understand exactly why we have been spared death from COVID-19. Yet, current daily life somehow is not allowing us to feel as though we have won this global battle. Why do our days feel covered in an invisible yet very palpable haze? I can answer that question...at least, partially.

As of this writing, we are into the 5th wave of COVID-19 infections within the United States. The FDA just recently approved the CDC's recommendation to authorize COVID-19 vaccinations (Pfizer-BionTech® and Moderna® brands only) for children as young as 6 months of age. Summer 2022 is staring us in our collective faces and we, as a society, are ready to venture out and vacation after more than 2 years of various restrictions. We want our lives back! And yet, we cannot seem to escape our preordained appointment with some universal truths of the natural world, including: (1) life exists independently of human beings and (2) life evolves on a trajectory and timeframe not dictated by human beings. In fact, all of life is subject to the immutable laws of nature.

As we resign ourselves to these inescapable truths, we begin to see the folly in trying to convince ourselves that we can simply declare a global pandemic "over", and it becomes so. Rather, we know in our innermost core that this cannot be truth. Instead, truth says that as members of a global society, we will likely continue to be subjected to infection, illness and death from various viral variants capable of causing COVID-19 disease. It is this inescapable fact that, in my opinion, keeps each of us living in what feels like that palpable haze. "Are we that weak," we ask ourselves, "as to not be able to triumph over a submicroscopic viral particle???" The answer is yes, we actually ARE that weak. Or more directly, evolution actually IS that powerful a force! I will leave it to you to come to an understanding as to how evolution came to be and continues to operate (hint: God).

And so, I will end this chapter by inviting you to consider the science of evolution. Try for now to put away any preconceived or previously taught ideas and opinions on the matter. Similarly, tuck away any negative biases you may currently possess. Now that we have removed those barriers, I'd like you to walk with me for just a short piece and consider: if evolution is a naturally occurring phenomena acting on all living things in unique ways prescribed in part by local microenvironments, then that means WE, as human beings, are also subject to evolutionary pressures. Taking this idea to a logical end concerning the children you are "honestly parenting", your children MUST be impacted by your efforts and approaches. All that you do, doing it from a place of love and trust, those efforts MUST materialize within the lives of the children you engage. That gift of Faith given freely to each of us eventually leads all of us back to its source. And that, my friends, is a humbling reality.

You and I must continue to choose to engage our children, question our children, listen to our children, challenge our children, protect our children, pray for our children, love our children and...ENJOY our children. Remember, it's not always the destination...it's the journey!

Have a blast!

Dr. Buckley

::: Epilogue :::
Keep Connected With Honest Parents

Wow. You made it to the end!
Allow me to be one of the first to say to you,
"I'm proud of you AND I'm excited for you!"

You had already made the choice to be present in your child's life before you picked up this book. In fact, it was your pure longing for the best life possible for your child that ultimately put this book on your radar. I cannot overstate just how awesome and faithful God is to have done this! You probably have questions triggered by various sections and topics we've covered. I want you to know that I have thought beyond the four corners of this book in that regard. I have setup a companion online portal for this book, chock full of practical tips and tools that will complement the themes that we've covered together. I encourage you to take advantage of these resources. There, you can download various worksheets to help you organize as you implement selected strategies from the book. Within the portal you can also connect with others, in the honest parenting community, to exchange ideas and encourage one another to stay committed. Take advantage of the opportunity to join our exclusive list of members who receive insider information on new book releases and book tour dates before the general public, as well as discounts on future workshops and online seminars where you and I can have more substantive interactions and exchanges.

This is just the beginning of our journey. In my "other world" of molecular biology we invoke the concept of "**cross-talk**". **Cross-talk** typically refers to instances where two or more different molecular proteins in different biochemical pathways are capable of interacting with each other such that they have an effect on biochemical pathways of which they are not members (I know...that was a mouthful!). As **honest parenting** becomes second nature to you and your family, you will begin to realize there is "cross-talk" within and among various parts of your lives. The honest parenting methods I have introduced to you in this book act as that "molecular protein", that "glue", if you will, that binds together what would otherwise be perceived as disconnected, disparate activities we all must attend to. These activities (raising children, maintaining and growing an enduring marriage, successfully navigating career milestones, balancing spiritual life with social and political realities, etc.) all come with their own peculiarities. But they all demand an honest approach in order to accomplish those goals that appear almost insurmountable.

A quick note here: some of my readers may have questions because the foundation from which I have taken you on this journey is rooted in Christianity. It is not lost on me that I have readers spanning the human experience; outside of the Christian worldview. If you are that reader, this section is for you. So, let me briefly deal with the following: "How is **honest parenting** accomplished outside the Christian paradigm?"

Notice that the question is *not* CAN it be done; the question is HOW to do it. I am confident that the principles identified and discussed throughout this book are accessible to you no matter your faith tradition. How can I make such a blanket and bold assertion? Well, think just a minute about what religion truly means to you. In doing so, you will discover that the concept of religion is more difficult to formally define than generally appreciated. A majority of scholars and theologians who have approached this exercise agree that a satisfying, universal definition may never be able to be articulated in written form. In fact, language itself can serve as a barrier. It is not unusual to find that the basic concept of religion, as we have come to know it in the Western world, does not even have corresponding words in other languages! Yet, the constructs that permeate and support all religious frameworks are rooted in the universality and commonality of humanity. Since humanity, humanness, being humane is experiential by design, it follows naturally that all societies engaged in the human experience will adopt universal Truth within their unique worldviews. Therefore, no matter your faith background, no matter your current religious practice, no matter the name of god you invoke, honest parenting principles definitely have a home in your world. I am confident of it.

I then leave it to you to be faithful (faithfulness is a must) in your personal walk with your god and allow spiritual revelation to weave together Truth and pour that Truth into you and your children because parenting is a BIG DEAL in and of itself; and, embracing this job as an **HONEST** parent actually might trigger the uninitiated to accuse us all of being a bit neurotic (i.e. you should be ready for concerns from family, co-workers and friends that you are taking your parenting much too "seriously," in other words)!

Quite frankly, on the surface, *Honest Parenting* does look like we're in hyperdrive with all its observation, questioning, analyzing, implementing, and modifying!!! And, perhaps the uninitiated are right to be curious about us being noticeably deliberate and driven, especially given that we're still "on our game", handling the day-to-day basics of parenting like cooking, cleaning, providing shelter, etc. So, yeah, I did mean it when I said that honest parenting is not for wimps, chumps or scary cats!

Now, on to some really good news—the secret behind **honest parenting**. The secret is this:

> **None of us can ever TRULY be everything that our children really need.**

I know, I know...just stay with me, please. All of humanity is made in the image of God; and, regardless of how "religious" any one of us may consider ourselves to be, all children are, by God's own account, the closest expressions we have of God's true Love in action. By that same Divine Designer, children are also humanity's most vulnerable treasures. Their safety and protection drive us (their parents) in our individual desires to use everything under our control, as human beings, to extend that much deserved and needed protection to them. In doing so, we encounter our own limitations in ensuring that our cherished children never see the deep fears we have of failing them or losing them. Truth be told, the only feeling we have that equals our adoration are our fears of failing them in one way or another.

I know what you're thinking right now..."Dr. Buckley, you said you had some 'really good news' what gives?" I'm getting to the good part, I promise. Hang in there. So, where was I? Oh yeah, how we all have these paralyzing fears (LOL).

So, we all have these paralyzing fears of not being able to protect them and/or of losing them. Even worse than these fears are the daily reminders of our natural limitations in our own ability to make manifest our deep love and need for protection. We yearn for those expressions toward our children to be perfect. Sadly, the hard truth is we will always fall short - because we are simple humans: frail, fearful, and lost. Make no mistake: we have a duty to be good stewards of all we have been given, including our children. But our humanity necessarily imposes limitations on our ability to provide the TRUE desires of our children and of all humanity: safety and acceptance; in a word, love. Not man's love, but true love. Once an **honest parent** accepts and embraces this stark reality, all of those self-sabotaging habits and emotions will begin to take their rightful place: in the pits of hell.

We can lift our hearts and our eyes and see that while honest parenting does take effort, we are not in this alone. The Father of Lights keeps and protects even us as we yield. Then and only then will we begin to sense the lifting of that burdensomely heavy weight of "I must be a perfect parent. I must produce a perfectly raised child. I must get this right and I have all the power to do so." No, you don't...no WE don't. But, it's ok. Thank God, we get to just be parents, and love and be loved by our children. Hallelujah! Amen.

Finally, here is the Secret (the capital is intentional; you know where I'm going with this, right? Full circle.)

Those challenging situations you will find yourself in the midst of as you parent honestly can actually become a lot easier to handle if you remind yourself to just have fun!

Yes, raising children is serious business. But, don't become so serious about and focused on the outcome that you miss the process. Don't forget to live with and be present for your children. It is in those moments that God seals the promises that become the desires of your heart. I pray this book has been of help to you. I encourage you to practice approaching life honestly each and every day. What will keep us all on this honest journey?

Our children deserve the type of world you and I have been challenged to help create for them. We know what it looks like, how it feels, how it should behave. Our challenge is to stay the course and remain focused on the goal. Now I need you to have fun during this process...considering you'll be applying and reapplying these (and other) principles throughout your life and the lives of your children. It makes no sense to do all of this work and not enjoy yourself in the process; and, I invite everyone one of you to read future titles in my *A Scientist's Approach*™ series so we can all continue this journey of discovery together.

Hope to see you all as I do my part in **Honest Parenting**!

Dr. Buckley

A SCIENTIST'S
APPR()ACH

A Scientist's Approach™ is a virological consultancy that provides accurate, time-sensitive solutions using multi-faceted, data-driven approaches. The firm was established to show the intersections of science with parenting, co-parenting, leadership, service, partnerships and how to apply faith-based virological solutions to COVID-19 in your home, business, or office.

Services available:

- Virological Consulting
- STEM Consulting
- Speaking

- Article Writing/Blogging
- Panelist
- Podcast Guest

For media, speaking and consulting inquiries, email info@ascientistsapproach.org.

Follow me on Facebook® and Instagram® @ ascientistsapproach

ascientistsapproach.org

INDEX

Behavior Modeling, 31–43, 81

Behavioral Modification, 31-43

Bullying, 63, 64, 81-83

Conflict Resolution, 32-34, 63, 64, 81-83

COVID-19, 38-45, 91-95

Dependent Variables, 25, 58

Feedback Loop, 45, 47

Hypotheses, 24, 25

Independent Variables, 25, 58

Ordered Process, 13-15, 18, 25

Reductionist Method, 14, 32, 61

Scientific Method, 31

Spiritual Process, 21-23

Transitive Property of Equality, 12

Visualization, 37-47

www.ingramcontent.com/pod-product-compliance
Lightning Source LLC
Chambersburg PA
CBHW072042040426
42447CB00012BB/2976